Erik M. Galimov · Anton M. Krivtsov
Origin of the Moon. New Concept

Erik M. Galimov
Anton M. Krivtsov

Origin of the Moon.
New Concept

Geochemistry and Dynamics

De Gruyter

Mathematics Subject Classification 2010: 96.20.Br; 96.20Dt: 95.10.Ce: 96.30.Bc; 29.25.Rm; 91.80.Hj; 98.80.Ft..

ISBN 978-3-11-028628-1
e-ISBN 978-3-11-028640-3

Library of Congress Cataloging-in-Publication Data

A CIP catalog record for this book has been applied for at the Library of Congress.

Bibliographic information published by the Deutsche Nationalbibliothek

The Deutsche Nationalbibliothek lists this publication in the Deutsche Nationalbibliografie; detailed bibliographic data are available in the internet at http://dnb.dnb.de.

© 2012 Walter de Gruyter GmbH & Co. KG, Berlin/Boston

Typesetting: PTP-Berlin Protago-TEX-Production GmbH, www.ptp-berlin.eu
Printing and binding: Hubert & Co. GmbH & Co. KG, Göttingen
♾ Printed on acid-free paper

Printed in Germany

www.degruyter.com

Contents

Introduction ... ix

I Geochemistry

1 The Moon as a celestial body ... 3
- 1.1 Size, mass, density ... 3
- 1.2 Moment of inertia ... 3
- 1.3 Orbital motion ... 4
- 1.4 Obliquities and inclinations ... 5
- 1.5 Angular momentum ... 6
- 1.6 Orbital evolution ... 7
- 1.7 Libration points ... 8

2 The history of the study of the Moon ... 10

3 The Moon as a geological body ... 13
- 3.1 Lunar gravity ... 13
- 3.2 Asymmetry of the lunar shape ... 14
- 3.3 Magnetic field ... 15
- 3.4 Topography ... 15
- 3.5 Lunar Rocks ... 19
  - 3.5.1 Highland rocks ... 20
  - 3.5.2 Maria rocks ... 22
- 3.6 Lunar chronology ... 23
- 3.7 Internal structure and temperature ... 25

4 Similarity and difference in composition of Earth and Moon ... 29
- 4.1 Iron content ... 29
- 4.2 Redox state ... 30
- 4.3 Volatiles ... 31
- 4.4 Refractory elements ... 33
- 4.5 Thermal evolution ... 35
- 4.6 Isotopic composition similarity ... 36
  - 4.6.1 Oxygen ... 36

		4.6.2	Silicon	39
		4.6.3	Titanium	39
		4.6.4	Cromium	40
		4.6.5	Tangsten	40
		4.6.6	Magnesium	41
		4.6.7	Litium	41

5 Hypotheses on the origin of the Moon   42

    5.1 Early hypotheses 42

    5.2 Giant impact concept and its weaknesses 44

6 The model of evaporative accretion   47

    6.1 Two possible paths of evolution of the solar nebula 47

    6.2 Introduction to the dynamic model 49

    6.3 Loss of iron and enrichment in refractories 51

    6.4 Asymmetry of accumulation of Earth and Moon 55

7 Geochemical constraints and how the giant impact and evaporative accretion concepts satisfy them   60

    7.1 Identity of isotope compositions of the Earth and the Moon 60

    7.2 Loss of volatiles without isotope fractionation 61

    7.3 Water in the Moon. D/H ratio 65

    7.4 Siderophile elements 70

    7.5 Constraints following from Hf–W systematic 79

    7.6 $^{129}I-^{129}Xe$ and $^{244}Pu-^{136}Xe$ 81

    7.7 U–Pb system. Time of accomplishment of accretion 82

    7.8 Rb-Sr system. Time of Moon origin 85

II Dynamics

8 Dynamical modeling of fragmentation of the gas-dust cloud   93

    8.1 Computational modeling using particle dynamics 93

    8.2 Computational technique 98

       8.2.1 Classical Barnes-Hut algorithm 98

       8.2.2 Data structure 101

       8.2.3 Tree traversal 103

       8.2.4 Parallelization scheme 104

       8.2.5 Comparative analysis of the performance 106

       8.2.6 Results of the computations, 2D-model 108

    8.3 Evaporation of the particles as an important factor of fragmentation .. 111

    8.4 3D-model of evaporative fragmentation 113

       8.4.1 Simulation parameters and numerical experiments 113

		8.4.2	Modification of the parameters (interaction potential, angular and random velocities, and dissipation coefficients) 116

                8.4.3 Variation of number of particles 121

                8.4.4 General trends in the system behavior 123

9 Dynamic modeling of accretion   125

    9.1 Computational model 125

    9.2 Determination of sizes of Earth and Moon embryos 135

    9.3 Consideration of precollapse evolution of the gas-dust cloud 137

    9.4 Temperature evolution 139

Conclusions   143

References   145

Index   163

Introduction

Sometimes interest in the problem of the Moon's genesis seems exaggerated. Why does this small celestial body, one of many in the solar system, attract so much attention?

First of all, the Moon genesis is part of the Earth genesis problem. Knowledge of our own planet, understanding of how and when its oceans and atmosphere came into existence, how and when the crust and the core of the Earth were formed, and how life originated on the Earth are not just academic issues; these are profound issues of human self-consciousness. Understanding the Earth's genesis is impossible without solving simultaneously the Moon genesis problem. The Earth and the Moon form a genetically linked pair. Moreover, a lot of questions relating to the early history of the Earth cannot be answered by studying the Earth alone. Although the age of the Earth is approximately 4.56 billion years, the oldest rocks ever found on Earth are no more than 4 billion years old, i.e., we have no material evidence of the initial 500–600 million years of Earth's history (except for a few zircon crystals). But the age of some rocks found on the Moon and delivered to Earth is $4.4 \sim 4.5$ billion years, and even older rocks may be found yet.

The Moon genesis may turn out to be central to the question of how the planets in the solar system have been formed.

In certain respects the Earth and the Moon form a unique pair. Among the inner planets of the solar system only Earth has a large satellite. Mercury and Venus have no satellites. Mars has two small, irregular-shaped satellites, Phobos and Deimos, the larger of which (Phobos) is 22 km longwise, while the Moon's diameter exceeds 3,500 km.

In relation to its planet, too, the Moon is the largest satellite in the solar system. The size ratio between the Moon and the Earth is 1:81.3. The size ratio between Ganymede, Jupiter's largest satellite, and Jupiter is 1:12,200. Moreover, the total angular momentum of the Earth-Moon system is many times higher than that of the other terrestrial planets.

Explanation of these peculiarities within the general planet accumulation theory is of critical importance in order to understand the laws of genesis and evolution of the solar System.

The most interesting and intricate point of the Moon genesis problem is the existence, on the one hand, of unique similarities in some respects of the composition of Earth and Moon – for example, isotopic composition of certain elements – and on the other hand, of fundamental distinctions – for example, as regards iron content and refractory elements. Hypotheses based upon the shared origin of the Earth and the Moon therefore need to explain the observed distinctions, whereas hypotheses allowing for independent origins of Earth and Moon materials have to explain the similarities.

A vast number of publications have been devoted to studies of the Moon, including the issue of its genesis. There are quite detailed reviews among them. We will provide references to these publications. At the same time, however, it would be unrealistic to expect readers, parallel to perusing this book, to read all the literature we may refer to. Our book therefore contains a necessary minimum of data about the Moon, that is, data immediately relevant to a discussion of the Moon genesis problem.

The major hypotheses advanced to date on the origin of the Moon include: 1) the capturing of the Moon by Earth; 2) the separation of the Moon from Earth due to rotational instability of the parent body; 3) the co-accretion of the Moon and Earth from a swarm of planetesimals; 4) the formation of the Moon as a result of a collision of Earth with a planet-sized body (the giant impact hypothesis).

Earth's capture of a body on its orbit is unlikely and is in contradiction with the isotopic identity of terrestrial and lunar material. The separation (fission) hypothesis has been abandoned for dynamical reasons: the total angular momentum of the Earth-Moon system is not sufficient for rotational instability, while attempts to redeem the situation by taking into account the resonance effects from the sun have failed, the relatively high dissipation resulting in damping of the resonance oscillations.

The co-accretion hypothesis failed to provide a plausible explanation of the Moon's iron deficit and simultaneously its enrichment in refractory elements. Currently most research is inclined to accept the giant impact model. According to this hypothesis collision of the Earth with another planet-sized body resulted in ejection to a near-Earth orbit of molten material, which then aggregated to produce the Moon. The impact hypothesis has good dynamical grounding. It provides a simple solution for the problem of iron deficit in the Moon, and plausible explanation of the high angular momentum of the Earth-Moon system. However, it also has some significant flaws.

Our book is largely devoted to presenting our own concept. We have tried to consider in an unbiased manner the extent to which the suggested hypotheses, and first of all our model compared to the mega-impact model, satisfy the most critical geochemical and physical constraints on the Moon origin.

The suggested model concordantly explains: 1) the observed deficit of iron in the Moon, 2) its enrichment in refractory elements; 3) the loss of volatiles by the Moon without isotope fractionation, 4) the formation of the double Earth-Moon system from a common source of primordial (chondritic) composition; 5) the identical isotope composition – $^{16}O/^{17}O/^{18}O$, Ti, W, etc. – of Earth and Moon; 6) the presence on the Moon of water and its isotopic composition; 7) the distribution of siderophile elements; we show 8) compliance of our model with the observed behavior of isotopic systems $^{182}Hf-^{184}W$, $^{87}Rb-^{86}Sr$, $^{129}J-^{129}Xe$, $^{244}Pu-^{136}Xe$, $^{238}U-^{206}Pb$, $^{235}U-^{207}Pb$; and within our model we offer 9) an interpretation of the high angular momentum of the Earth-Moon system.

If our hypothesis of the Moon genesis is valid, it may become the basis for a brand new approach to the problem of planet accumulation. Our hypothesis is inconsistent with the generally accepted theory of accumulation of planetary bodies in the solar system. This is its weak point, although it may eventually prove to be an achievement. We claim that evolution of the solar nebula includes formation of gas-dust clumps (not only solid planetesimals), which by non-elastic collision may have accumulated into a large (within the Hill sphere) gas-dust body, the initial stage for planets. Gradual compression of this body and its fragmentation (subject to certain conditions) resulted in the formation of embryos of the Moon and the Earth. We consider geochemical and dynamical consequences of this mechanism and find satisfactory consensus with the observed facts. But it is not our purpose here to present a general evolutionary model of the protoplanetary nebula. We regard consideration of a new model of accumulation of planets as a separate issue and leave this to be resolved in the future.

Part I
Geochemistry

Chapter 1

The Moon as a celestial body

1.1 Size, mass, density

The mass of the Moon is $7.351 \cdot 10^{25}$ g (Table 1.1). Its shape is close to spherical, the radius is 1,737.4 km. The lunar sphere volume is $21.99 \cdot 10^9$ km^3. Hence, lunar density is 3.343 g/cm^3. Such density is typical of rocks such as eclogite and peridotite.

Table 1.1. The Moon as compared with the Earth.

Mass of the Earth	$5.977 \cdot 10^{27}$ g
Mass of the Moon	$7.351 \cdot 10^{25}$ g
Density of the Earth	5.517 g/cm^3
Density of the Moon	3.343 g/cm^3
Radius of the Moon	1,737.4 km
Surface area of the Moon	$37.96 \cdot 10^6$ km^2
Lunar sphere volume	$21.99 \cdot 10^9$ km^3
Moon vs. Earth weight ratio	1/81.3

The mass of the Moon is ∼ 80 times lower than the mass of the Earth. However, the Moon is the heaviest satellite in the solar system in relation to its central planet. The largest satellite in the solar system is Ganymede, but it weighs 12,200 times less than its planet, Jupiter.

The density of the Moon is much lower than that of the Earth. This is partially explained by the massiveness of the Earth and by the fact that the matter inside it is compressed under high pressure. But even if one considers the densities of the Earth and the Moon under the same normal pressure, the density of the Moon appears to be lower than that of the Earth: 3.34 and 4.54 g/cm^3, respectively. This means that the Moon contains much less iron than the Earth. The Earth has a heavy core accounting for ∼ 32 % of its mass. Considering the average known density of the Moon, it can only have a small metallic core. This corresponds with the limitation that is imposed on the core size by the value of moment of inertia.

1.2 Moment of inertia

The moment of inertia is a measure of resistance of a body to change in its rotation. The lunar solid moment of inertia I_{solid} is determined by use of accurate Lunar Laser

Ranging. The measurements of the distance from the land-based observatories to the laser reflectors on the Moon's surface provide parameters for evaluation of the moment of inertia (Konopliv et al., 2001). The most recently obtained value of dimensionless moment of inertia $I_{\text{solid}}/MR^2 = 0.3930 \pm 0.0003$ (Williams et al., 2012), where M and R are the mass and radius respectively.

Homogeneous solid sphere (ball) has $I_{\text{solid}}/MR^2 = 0.4$. The value 0.393 means that there is increase of density with depth, but the form of its increase is not uniquely determinable. Calculations show that given the observed values of the moment of inertia and density the **Moon cannot contain a metallic core more massive than 5 % of the total lunar mass** (Hood, 1986).

The Earth has a lower dimensionless moment of inertia ~ 0.330 since it has significant concentration of density in the central part of the body in the form of a large metallic core.

Accurate Lunar Laser Ranging provided the first evidence for a fluid lunar core (Williams et al., 2001). Variation in rotation and orientation of the Moon is sensitive, among other parameters, to dissipation due to relative motion at the fluid-core/solid mantle boundary.

1.3 Orbital motion

The Moon rotates around the Earth at a distance of 384,400 kilometers. To be precise, the Moon and the Earth rotate around their common barycenter, which is situated inside the Earth 4,670 km from its center. Eccentricity of the lunar orbit ranges from 0.0435 to 0.0715 (the average value is 0.0555).

The period during which the Moon completes one turn around the Earth is known as the siderical month. It lasts 27.32166 earth-days. With regard to the Earth's own orbital movement around the Sun, the period between similar lunar phases appears equal to 29.53059 earth-days, and is called 'the sinodic month'.

The average orbital travel speed of the Moon is 1.023 km/sec.

The Moon rotates in prograde motion, that is, in the same direction as the Earth rotates around the Sun.

The Moon's rotation has been synchronously locked by the Earth-Moon tidal interaction. Therefore the same hemisphere of the Moon (near side) always faces the Earth.

The prograde rotation *per se* requires explanation (Dones and Tremaine, 1993).

The planets are believed to have accreted most of their angular momentum from solid planetesimals. The Keplerian motion of planetesimals in a disk suggests that the part of the planet closest to the Sun should move the fastest, which implies retrograde rotation (Lissauer et al., 2000).

It follows from the above that purely solid-state accumulation of planets by means of their growth via accumulation of planetesimals would lead to retrograde rotation of the planets.

However, six of the eight planets in the solar system rotate around their axis in the same direction in which they revolve around the Sun (prograde rotation). Venus rotates very slowly in retrograde direction, and the axis of rotation of Uranus and its satellite system lies almost on the ecliptic plane.

In contrast, viewed as a fluid a Keplerian disk has positive vorticity, suggesting that planets should rotate in the prograde direction.

When any instability occurs in the gas and dust protoplanetary nebula surrounding the Sun, and such instability causes formation of vortices, the resulting gas and dust aggregation will rotate in the prograde direction. As the aggregations collide with each other, their moment values will be added together. If planets are formed via accumulation of the aggregations, then the ultimate body will rotate in the prograde direction.

It follows that **accumulation of planets must, one way or another, include the accumulation phase of the gas and dust aggregations.**

The angular momentum of a planet can be modified by collisions with solid bodies and is finally determined by these two accumulation mechanisms.

1.4   Obliquities and inclinations

Obliquity is the angle between rotational and orbital angular momentum vectors (Fig. 1.1). The equatorial plane of the Earth is tilted relative to the ecliptic plane (the orbital plane of the Earth) at an angle of 23.4°, but that of the Moon is tilted at an angle of only 1.5° (1° 32' · 47''). Inclination of the lunar orbit to the ecliptic plane is about 5° (varies from 4° 59' to 5° 17').

Thus there exists a significant angular difference between Earth's equatorial plane and the lunar orbit plane. However, it cannot be considered as a constraint on lunar origin, since it cannot be excluded that the present angular difference between inclination of the lunar orbit and the obliquity of the Earth's rotation axis was established at an early stage by the impact of a single relatively small planetesimal at high latitude on Earth or on the Moon (Wood, 1986; Boss and Peale, 1986).

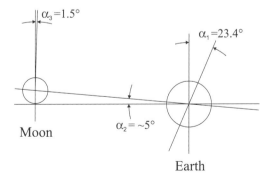

Figure 1.1. Obliquities in the Earth-Moon system: α_1) obliquity of the Earth's axis; α_2) inclination of lunar orbit; α_3) obliquity of the lunar axis.

Observed inclinations of rotation axes of the Earth and the Moon can be caused by impacts of relatively small bodies, whose weight does not exceed $10^{-2} - 10^{-3}$ of the planetary body mass (Safronov, 1969).

Moreover the planet's obliquity can change as a result of changes in the orientation of either its spin axis or its orbit pole. "Planetary rotation can be altered subsequent to accretion, so spins may have been more randomly oriented in the solar system's history" (Lissauer et al., 2000). The obliquities cannot be considered primordial since they changed during chaotic variations caused by spin-orbit resonances in the solar system (Laskar and Robutel, 1993). **Therefore obliquities and inclination of the Earth and Moon axes cannot serve as a constraint on planet origin.**

1.5 Angular momentum

The amount of angular momentum in the Earth-Moon system is $3.45 \cdot 10^{41}$ rad $\cdot$ g $\cdot$ cm^2/s. This value is the sum of the rotational angular momentum of the Earth and the orbital angular momentum of the Moon.

Some believe that the Earth-Moon system has anomalously high rotational angular momentum. Indeed, its magnitude is highest among terrestrial planets. This fact is considered as evidence of a mega-impact event that resulted in additional angular momentum of the proto-Earth and the impactor.

However, **the total angular momentum of the Earth and the Moon cannot be considered anomalous**. It is consistent with the relationship between different cosmic bodies starting from small asteroids and ending with stars, which have a more or less similar rotation period close to their rotational instability (Fig. 1.2).

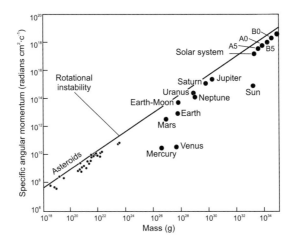

Figure 1.2. Angular momentum density as a function of mass for the solar system bodies and stars of some classes for comparison (after Ringwood, 1979).

The maximum rate at which a body is able to retain material (loosely bound) at its equator can be calculated by setting the centrifugal force equal to the gravitational force. For a spherical body of density ρ, this yields a minimum rotational period (Lissauer et al., 2000)

$$T_{min} = \sqrt{\frac{3\pi}{G\rho}}$$

where G is the gravitational constant.

It is seen from the equation above that minimum rotation period (maximum rotation rate) does not depend on the size of the body.

The magnitudes of the rotational periods on Fig. 1.2 vary within one order while variations in mass span 14 orders.

Small cohesive bodies like asteroids may rotate faster than the limit defined by the equation for T_{min}. However, they conform with the general relationship and rotate near that limit. The giant planets of the solar system rotate at a rate that is only half the limit of rotational stability. The same is characteristic of the Earth-Moon system. If the orbital angular momentum of the Moon were added to the Earth's spin, it would rotate at approximately half the rate required for breakup.

The sun itself and the terrestrial planets Mercury, Venus, Mars (and the Earth without the Moon) rotate at less than one-tenth breakup.

The above regularity appears to be quite universal if one considers the wide range of the space bodies it applies to, and also appears to comply with the physical approach, since it is collinear with the dependence between angular momentum values responsible for the rotational instability of such bodies (Ringwood, 1979).

From this standpoint, **it would be logical to consider the value of specific angular momentum of the Earth-Moon system as the normal value, while considering the lower momentum values of the other planets within the inner area of the solar system as abnormal values caused by a partial loss of angular momentum that took place at a certain time in their history.**

1.6 Orbital evolution

At present, the orbital angular momentum of the Moon is about 5 times larger than the spin angular momentum of the Earth. Significant orbital evolution can be attributed to tides.

There is an exchange of angular momentum between the Earth and the Moon due to tidal interactions. The Moon was once much closer to the Earth than it is now. The tidal interactions have decelerated the rotation of the Earth while accelerating the Moon's and expanding the lunar orbit (Burns, 1977).

The angular differences between orbital elements of the Earth and the Moon can have occurred during the orbital evolution (Wood, 1986; Boss and Peale, 1986).

1.7 Libration points

There are a number of points in the space around the Earth, in the common gravitation field of the Earth, the Moon and the Sun, where the resultant of the forces applied to a body placed in that point is close to zero. These are libration (Lagrangian) points.

There are five points of this type (Fig. 1.3). Three of them are situated on the straight line between the centers of the Earth and the Moon; these are collinear libration points L_1, L_2 and L_3. The other two points (L_4 and L_5) are located in the corners of equilateral triangles opposite the base connecting the centers of the Earth and the Moon.

The position of such points does not change in relation to the Moon and the Earth as they travel along their orbits. A body with negligible weight compared to the Earth and the Moon may stay in those points without being subjected to any displacing force.

For example, if an object (satellite) rotates around the Earth in an orbit closer than the Moon, it must in accordance with Kepler's laws move faster than the Moon. Therefore the distance between the Moon and the third object continuously changes. However, there is a point on the line between the Earth and the Moon (L_1 in this example) where due to the Moon's gravity the attraction from the Earth becomes weaker and the object needs less speed to maintain its orbit. At the Lagrangian point its angular speed is exactly equal to the angular speed of the Moon. Therefore as they rotate all three objects stay in line where the Moon and the third object have equal angular velocities. A similar type of balance exists at other Lagrangian points.

Moreover, the position of such objects at points L_4 and L_5 is stable, i.e., if any external factors (collisions, etc.) shift the object from its position at the libration point,

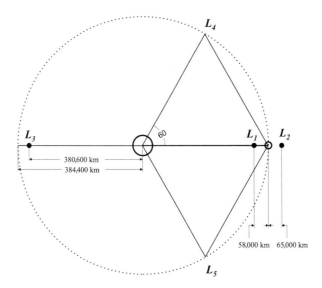

Figure 1.3. Libration points in the Earth-Moon system. L_1, L_2, L_3 – collinear libration points; L_4, L_5 – triangular libration points.

the object will try to return to that position. Therefore, matter may accumulate in these points. The libration points L_4 and L_5 are gravitation traps. Indeed, dust concentration is observed in these points. However, this concentration is negligible. Only sites centered on Jupiter's stable triangular Lagrangian points contain some debris (Trojans).

Gravitational features of the libration points have given rise to one version of the mega-impact hypothesis (Belbruno and Gott, 2005). As is known, isotopic composition of the Earth and the Moon is similar with respect to certain elements, in particular oxygen, which is evidence of genetic kinship between the Moon and the Earth. At the same time the dynamic mega-impact model indicates that if such an event had actually taken place, the Moon would consist primarily of the impactor material, i.e., of the material that has no cosmic relation to the Earth. This contradiction would be eliminated if the Earth and the impactor had been formed in adjacent areas of space – for example, if the impactor had emerged in one of the libration points (L_4 or L_5) within the Earth-Sun system. But such a scenario is unlikely since the mass of the impactor in the mega-impact hypothesis must be comparable with that of the Earth, whereas stable accumulation in the libration points is possible for a body or an aggregation whose mass is small compared to the mass of the Earth. Even if it would happen the speed of the collision is too low to provide the scenario of the megaimpact origin of the Moon (see e.g. Cameron, 1997).

Chapter 2

The history of the study of the Moon

Scientific study of the Moon started when a telescope was trained on it for the first time. The Dutchman Hans Lippershey invented a primitive telescope at the beginning of the 17th century. Two years later, in 1610, Galileo Galilei improved its scope and began systematic observations of the Moon. He discovered the lunar craters, noted highlands and lowlands, and established that the dark spots seen by the naked eye are extensive lowland areas. Then Galileo discovered the 4 satellites of Jupiter. It followed that the Moon was not unique, and the circulation of satellites around a planet was a common feature of the solar system.

By the middle of the 17th century the Italian astronomer Jambatista Riccioli had given names to more than two hundred formations observed on the Moon. In particular, he gave one of the most beautiful lunar craters the name "Copernicus", a dangerous move in those days. Just before, in 1600, Jordano Bruno had been burnt at the stake, and in 1630 the Inquisition forced Galileo to recant the heliocentric theory.

By the second half of the 20th century a photographic map of the Moon had been produced, containing details of more than 40,000 features. It had been found that the radius of the Moon was a quarter of the Earth's radius and its mass one eightieth of the Earth's mass. The density of the Moon, its moment of inertia and astronomical parameters of its orbit had been precisely measured.

A new epoch began in 1959. On January 2, 1959 the soviet spacecraft Luna-1 was launched toward the Moon. The spacecraft missed the Moon by 6,000 km. In the same year Luna-2 hit the Moon. (Table 2.1) Thus the first man-made object reached the lunar surface. The following year Luna-3 took the first photographs of the lunar far side, which humans had never seen before. A further significant achievement came when Luna-9 made the first soft landing and shot a TV panorama of the lunar surface. Before the Luna-9 landing there was no evidence that the lunar surface would support the spacecraft or be solid enough to walk on. Then the first humans orbited the Moon in Apollo-8 and looked back at the Earth for the first from the lunar distance. And in July 20, 1968 Apollo-11 landed the first humans on the Moon's surface, the apotheosis of lunar exploration. It was followed by five successful American human landings by Apollo missions. The astronauts made some significant measurements (magnetic field, heat flow, seismic) and collected more than 300 kilos of lunar samples, which were brought back to Earth's laboratories for study.

There were also three successful automatic sample return missions by the soviet spacecrafts Luna-16, Luna-20 and Luna-24. Although they returned less than 300 g of lunar material all told, theirs was a great engineering and scientific achievement. They

Table 2.1. History of lunar exploration*.

Launch date	Mission	Accomplishment
2 Jan 1959	**Luna 1**	**FIRST lunar flyby, magnetic field**
12 Sept 1959	**Luna 2**	**FIRST hard landing, magnetic field**
20 Apr 1960	**Luna 3**	**FIRST photos of lunar far side**
26 Jan 1962	Ranger 3	Missed the Moon by 36,793 km
23 Apr 1962	Ranger 4	Crashed on the lunar far side
18 Oct 1962	Ranger 5	Missed the Moon by 724 km
2 Apr 1963	**Luna 4**	**Missed the Moon by 8,500 km**
30 Jan 1964	Ranger 6	Hard landing, television failed
29 July 1964	Ranger 7	Hard landing, close-up TV
17 Feb 1965	Ranger 8	Hard landing, close-up TV
21 Mar 1965	Ranger 9	Hard landing, close-up TV
9 May 1965	**Luna 5**	**Crashed on the Moon**
8 June 1965	**Luna 6**	**Missed the Moon by 161,000 km**
18 July 1965	**Zond 3**	**Photographed lunar far side**
4 Oct 1965	**Luna 7**	**Crashed on the Moon**
3 Dec 1965	**Luna 8**	**Crashed on the Moon**
31 Jan 1966	**Luna 9**	**FIRST soft landing, TV panorama**
31 Mar 1966	**Luna 10**	**FIRST lunar.satellite, gamma-spectra, magnetic and gravity measurements**
30 May 1966	Surveyor 1	On-surface TV, soil- mechanics
10 Aug 1966	Lunar Orb 1	TV imaging, radiation, micrometeorites
24 Aug 1966	**Luna 11**	**Gravity, micrometeorites**
22 Oct 1966	**Luna 12**	**TV imaging from orbit**
6 Nov 1 966	Lunar Orb 2	TV imaging, radiation, micrometeorites
21 Dec 1966	**Luna 13**	**On-surface TV, soil mechanics**
5 Feb 1967	Lunar Orb 3	TV imaging, radiation, micrometeorites
17 Apr 1967	Surveyor 3	On-surface TV, soil-mechanics
4 May 1967	Lunar Orb 4	TV imaging, radiation, micrometeorites
19 July 1967	Explorer 35	Fields and particles
Launch date	Mission	Accomplishment
1 Aug 1967	Lunar Orb 5	TV imaging, radiation, micrometeorites
8 Sept 1967	Surveyor 5	On-surface TV, first chemistry data
7 Nov 1967	Surveyor 6	On-surface TV, chemistry
7 Jan 1968	Surveyor 7	On-surface TV, chemistry
7 Apr 1968	**Luna 14**	**Gamma spectra, magnetic measurements**
14 Sep 1968	**Zond 5**	**FIRST lunar flyby and Earth return, returned biological objects and photos**
10 Nov 1968	**Zond 6**	**Lunar flyby and Earth return, returned biologicalobjects and photos**
21 Dec 1968	Apollo 8	FIRST humans to orbit the Moon
18 May 1969	Apollo 10	FIRST docking in lunar orbit

Table 2.1. *continued.*

13 July 1969	**Luna 15**	**Failed robot sampler**
16 July 1969	Apollo 11	FIRST humans on the Moon (20 July)
8 Aug 1969	**Zond 7**	**Lunar flyby and Earth return, returned biological objects, photos**
14 Nov 1969	Apollo 12	Human landing, Oceanus Procellarum
11 Apr 1970	Apollo 13	Aborted human landing
12 Sept 1970	**Luna 16**	**FIRST robot sample return, Mare Fecunditatis**
20 Oct 1970	**Zond 8**	**Lunar flyby and Earth return, returned photos, landing in the Indian Ocean**
10 Nov 1970	**Luna 17**	**FIRST robotic rover Lunokhod 1, Mare Imbrium**
31 Jan 1971	Apollo 14	Human landing, Fra Mauro
26 July 1971	Apollo 15	Human landing, Hadley-Apennine
2 Sept 1971	**Luna 18**	**Failed robot sampler**
28 Sept 1971	**Luna 19**	**Orbiter, lunar gravity, TV, micrometeorites**
14 Feb 1972	**Luna 20**	**Robot sample return, Apollonius**
16 Apr 1972	Apollo 16	Human landing, Descartes
7 Dec 1972	Apollo 17	Human landing, FIRST geologist on the Moon, Taurus-Littrow
8 Jan 1973	**Luna 21**	**Lunokhod 2, Le Monier**
29 May 1974	**Luna 22**	**Orbiter, lunar gravity, TV, micrometeorites**
28 Oct 1974	**Luna 23**	**Failed robot sampler**
9 Aug 1976	**Luna 24**	**Robot sample return, Mare Crisium**

* Soviet missions are shown in bold-italic, US missions are shown in normal font.

demonstrated the feasibility of fully automatic sampling and sample return. And they extended the Apollo collection, as the samples returned by the robotic Lunamissions were collected at different sites.

During the Luna-17 mission an automatic rover was used for the first time. The rovers Lunokhod-1 (1970) and Lunokhod-2 (1973) were driven several kilometers over the Moon's surface via remote control from Earth.

The last mission of that unprecedented lunar race was by Luna-24, which returned samples from Mare Crisium in August 1976.

The latest period of Moon study started with the Clementine mission in 1994, followed by the Lunar Prospector mission, the Japanese lunar satellite (with two sub-satellites) SELENE (Kaguya), the Indian lunar orbiter Chandrayan, the Chinese orbiter Chang-E and NASA's spacecraft Lunar Reconnaissance Orbiter. They have provided important new data on the presence of water in the Moon, high precision mapping of the gravitational field and lunar topography and global mapping of some elements.

For centuries the Moon was studied as a celestial body. For two decades in the 60s and 70s, plus in the last few years, the moon has been comprehended as a geological object.

Chapter 3

The Moon as a geological body

3.1 Lunar gravity

A gravitational field is described through the value of gravitation potential. Equal potential values form an equipotential plane. Equipotential planes are spherical, if the body is a solid ball. If mass distribution is not uniform, equipotential planes are deformed due to local excess or lack of mass (Fig. 3.1).

The modern method of gravitational field studies is registration of gravitation perturbations of artificial satellite orbits. The lunar gravitational field was studied for the first time by means of the Soviet satellite Luna-10 in 1966 (Akim, 1966). These studies were continued by a series of the US satellites Lunar Orbital (LO-I, LO-II, LO-III, LO-IV and LO-V), which were placed in orbit with various orbital inclination and eccentricities. They provided quite an accurate gravity map for that time. The map displayed large positive gravity anomalies within large circular maria basins with low topography (Muller and Sjogren, 1968). These features were called mascons (short for "mass concentrations").

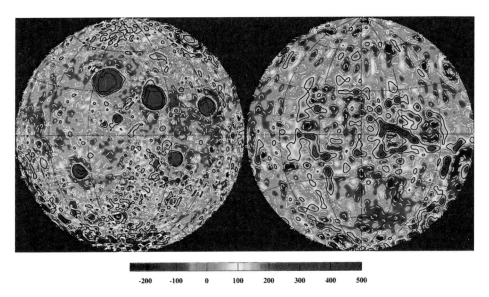

Figure 3.1. Vertical acceleration at the lunar surface for the LP 165P gravity field (from Konopliv et. al., 2001). The near side on the left, the far side on the right. Units are milligals with contour lines every 100 milligals.

Gravity data were significantly improved by the Clementine, Lunar Prospector (LP) and Selene missions. The Clementine laser altimetry data provided the global shape of the Moon for the first time (Smith et al., 1997).

Prior to LP, all known mascons were on the near side and associated with large maria-filled impact basins. Several new mascons were found for impact basins with little or no evidence of maria fill (Neumann et al., 1996).

Mascons are the most prominent features of the lunar gravity field. They are believed to be the result of a combination of a mantle plug and mare fill in the basin (see review in Konopliv et al., 1998). Each mascon anomaly has a significant contribution from the higher density mare relative to the crust (~ 3.3 vs. $\sim 2.9\,\text{gm/cm}^3$) as shown for Serenitatis by Phillips et al., (1972). Mascons have sharp shoulders with a gravity plateau and are surrounded by a negative gravity anomaly (Konopliv et al., 2001). Formation of a giant intrusion beneath a mare together with the volcanism apparent on the surface might have contributed to the creation of mascons (Mc Govern, 2012).

Using its two sub-satellites the Japanese Selene (Kaguya) mission produced the most accurate gravity map of the far side (Namiki et al., 2009). In contrast to the near side only a few maria basins were found on the far side.

3.2 Asymmetry of the lunar shape

The topographies of the near side and the far side of the Moon are different. The thickness of the crust is also different: the far-side crust is thicker than the near-side (Smith et al., 1997).

The prominent geological feature of the Moon is the terrain and elevation dichotomy.

Most of the maria are located on the near side while the far side has more highlands. In addition, the center of mass of the Moon is displaced to the near side by 2 km relative to the center of figure.

It has been suggested that the cause of the observed asymmetry is an internal mass redistribution (Parmentier et al., 2002). According to this theory, a dense Fe- and Ti-rich layer at the bottom of the magma ocean moved towards the hemisphere on the Earth-Moon line. This replacement could be the cause for the different thermal and volcanic history on the near side and the far side (Perera and Garrick-Bethell, 2011).

The topography dissimilarity of the near-side and far-side gravity is believed to be due to their different early thermal history. During bombardment 3.8–4 Ga (billion years) ago, when most of the Moon's impact topography was formed, the interior of the near side was still hot while on the far side it became cold. The softened interior material on the near side was easily distorted and compressed under impacts, producing mascons, whereas the crust on the far side was cooler and thus firmer.

Recently M. Jutzi and E. Asphang (2011) proposed that the observed asymmetry is due to late accretion by the Moon of a second satellite orbiting the Earth. The companion of the Moon was calculated to have a diameter of about 1,300 km and a speed of ~ 2 km/s relative to the Moon at the moment of collision.

However, Paul Warren (Warren, 2012) has shown that this model runs into serious chemical and physical difficulties.

3.3 Magnetic field

The Moon has no dipole field of its own. At the same time, there exist certain local magnetic anomalies emerging in areas that range from dozens to hundreds of square kilometers (Lin et al., 1998). Some magnetic anomalies on the back side of the Moon registered by the Lunar Prospector spacecraft appear to correlate with young, circular antipodal maria in the visible hemisphere (Hood et al., 1999). The nature of the local magnetic anomalies on the Moon is open for discussion. Suggestions have been made that they may have originated from piezoelectric effects caused by meteorite and comet impacts.

Samples of the oldest rocks delivered by the Apollo missions reveal residual magnetization, which justifies the suggestion that the Moon may have had its internal magnetic field 3.6–3.9 billion years ago (Cisowski and Fuller, 1986; Runcorn, 1996). According to Runcorn (1996), a weak lunar magnetic field existed prior to 4 Ga with an intensity of $\sim 4\,\mu T$ (4 microtesla or 0.04 Gauss in CGC units) with a sharp increase of intensity up to $100\,\mu T$ in the interval 4.1–3.9 Ga. However, Weiczorek et al. (2006) argue that the lunar core dynamo would not have been able to generate a magnetic field of $\sim 100\,\mu T$ at the lunar surface. Careful studies of an ilmenite basalt sample with a crystallization age of 3.6 Ga (sample 10,017) have shown that observed residual magnetization of the sample could not have been caused by any impact, and that it is connected with an external magnetic field (Suavet et al., 2012). The authors conclude that the Moon had an active dynamo at least between 4.2 and 3.6 Ga. Studies of some younger rocks (3.2–3.3 Ga) reveal no signs of paleomagnetism (Tikoo et al., 2012). The lunar dynamo ceased by 3.3 Ga and the core dynamo of Mars ceased at around 4 Ga (Lillis et al., 2008; Arkani-Hamed, 2012).

3.4 Topography

The key elements of lunar topography are lunar continents and maria. The continents are highlands featuring contrasting mountainous relief. Visually these are light areas on the lunar surface. The maria are lowlands, seen as dark spots on the lunar disk (Fig. 3.2).

The continents occupy over 80 % of the lunar surface, mostly on the far side of the Moon, where they account for more than 95 % of the surface.

The highest elevation on the Moon is about 10,755 meters (Leibniz mountain), and the lowest is about 9,060 meters deep in the South Pole Aitken Basin. Both are on the far side.

Continental rocks consist of magmatic series, in which ferroan anorthosites prevail. They have originated from early melting of the lunar crust, and are the oldest rocks.

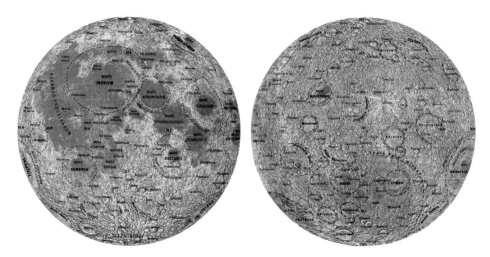

Figure 3.2. From atlas Antonia Rüke (2004).

During the subsequent history of the Moon they were subjected to meteorite impacts, some of which were from large bodies. As a result, to a depth of 20–25 km they consist of breccias, i.e., fragments of impact fractioning cemented by impact melts.

The maria occupy around 20 % of the surface. Almost 90 % of these are located on the visible side of the Moon. They form the bottom of gigantic craters filled with basalt lava. The maria surface features craters and volcanic structures, furrows and traces of lava flows.

Superimposed relief of the lunar surface is almost always characterized by craters. Crater sizes vary from gigantic craters forming large basins to microscopic structures.

The South Pole Aitken (SPA) basin is the largest impact structure on the Moon, 2400 km in diameter and about 8 km deep. SPA is also probably the oldest structure. It has a severely eroded surface and abundant superimposed craters.

Mare Imbrium, with a diameter of about 1,200 km, is an example of a gigantic crater in combination with the formation of maria basins.

As craters grow larger, their structure changes from a simpler cup-like type to more complex reliefs (Fig. 3.3).

Craters 10–15 km in diameter have a cup-like shape with level bottom and steep (up to 30–40°) slopes. The crater depth to diameter ratio is 0.2–0.25. Craters of larger diameter have a central hill and internal concentric circular elevations. Their bottom is flatter. The crater depth to diameter ratio of a 25–40 km crater is 0.1–0.15. The central hill may be as high as 1.5 km. A typical example of a central hill crater is crater Römer (Fig. 3.3) located on the continent between Mare Serenitatis and Mare Crisium (Legostaev and Lopota Eds., 2011).

Among craters with complex structure and high hills are such young and well-preserved craters as Tycho (85 km) and Copernicus (93 km).

Section 3.4 Topography 17

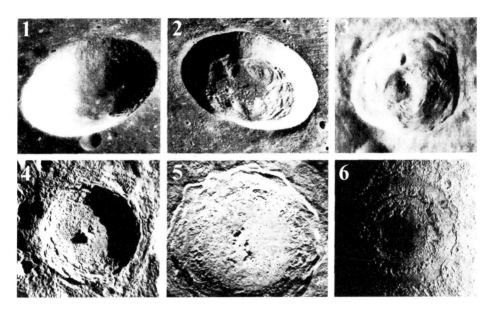

Figure 3.3. Transition from simple structure craters to more complex ones:
1 Isidor, a simple cup-like crater, diameter 15 km (coordinates: 4, 2° S, 34, 1° E), photo taken by Apollo-16 (NASA);
2 Bessel, a flat bottom crater, diameter 17 km (21, 8° N, 17, 9° E), photo taken by Apollo-15 (NASA);
3 Römer, a crater with central hill, diameter 39 km (25, 4° S, 36, 4° E), photo taken by the Lunar Orbiter-4 station (NASA);
4 Crater Tycho, diameter 85 km (43° S, 11° W), photo taken by the Lunar Orbiter-5 station (NASA);
5 Crater Copernicus, diameter 93 km (10° S, 20° W), photo taken by the Lunar Orbiter-4 station (NASA);
6 Multi-ring basin in Mare Orientale, diameter 900 km (20° S, 95° W), photo taken by the Lunar Orbiter-4 station (NASA) (from Legostaev and Lopota, 2011, by permission).

What are referred to as lunar mountain systems are usually the flank walls of large craters, and sometimes combined walls of superimposed impact structures.

Mare Orientale is a typical example of a multi-ring crater featuring four concentric edge walls with an outer diameter of 900 meters. It is named the Cordilleras system. The inner circular structures with diameters of 620 and 480 km are called the Rocky Mountains.

Lunar mountain systems known as the Carpathians, the Caucasus, the Apennines and the Alps form parts of the circular wall embracing the gigantic impact structure called Mare Imbrium.

Another important component of the relief is crater ejecta. A clearly visible system of crater ejecta rays from Copernicus extends almost 2,000 km.

Figure 3.4. Stone fields. A fragment of the video taken by Lunokhod-2. The cross dimension of the largest stone is ∼ 1 m.

The ejecta from large and small craters fill the lunar surface with stones varying from centimeters to many meters in size. Particularly large stones are concentrated around the crater walls.

Stones measuring from 20 cm to 2 m are typical of the lunar relief. Normally there are 2–4 stones per m^2 of the lunar surface (Fig. 3.4).

Precise global topographic data on the Moon with about 5 m accuracy was obtained by a laser altimeter on board the Kaguya spacecraft (Araki et al., 2009).

The lunar soil consists of a layer of dust-like, grainy substance known as regolith. Granulometric composition of regolith varies from 100 μm to a few millimeters. Its lithological composition depends on both regional rock mix and materials imported by impacts. Regolith takes shape under the influence of space particle radiation, solar winds and impacts of micro-meteorites.

The regolith surface is dotted with multiple cavities that form a typical honeycomb structure (Fig. 3.5).

Figure 3.5. Honeycomb structure of the lunar regolith surface. A fragment of the video taken by the Luna-9 station (from Legostaev and Lopota, 2011, by permission).

Impacts of micro-meteorites cause rock particles to break and fuse. As a result, agglutinates emerge, i.e., aggregates of the particles bonded by impact melts. The longer the time of exposition, the higher is the concentration of agglutinates in the soil. Protons from the solar winds promote reduction of FeO. Nano-sized zero-valency particles of metallic iron appear in the regolith. Maturity of the regolith is measured by the agglutinates and by concentration of $F°$ nano-particles.

The rate of accumulation of regolith is very low. It takes several hundred million years to accumulate a detectable two meter thick regolith layer (Sharpton and Head, 1982).

3.5 Lunar Rocks

The lunar continents and maria do not just differ in terms of topography; they also represent different petrological provinces composed of various rocks. Samples of lunar rocks were delivered to the Earth from 9 areas of the Moon by 6 manned Apollo missions and by 3 automatic Luna stations (Table 3.1). The lunar material collection has been further increased lately thanks to lunar meteorites.

Around 50 meteorites have been found in the Antarctic ice and in deserts and identified as having lunar origins. They were knocked out of the lunar surface by meteorite

Table 3.1. Locations of sites where samples were collected by Apollo and Luna missions.

Mission	Date	Location	Landing site	Type of rock
Apollo-11	21.07.69	0°41′N 23°26′E	Mare Tranquillitatis	High-Ti mare basalts
Apollo-12	19.11.69	3°11′S 23°23′W	Oceanus Procellarium	Low-Ti, low-Al Mare basalts, alkali basalts
Luna-16	12.09.70	0°41′S 56°18′E	Mare Fecunditatis	Low-Ti, High-Al basalts
Apollo-14	5.02.71	3°40′S 17°28′W	Fra Mauro, Oceanus Procellarium	Low-Ti mare basalts, KREEP alkali basalts
Apollo-15	31.07.71	26°26′N 3°39′E	Mare Imbrium	Low-Ti, Low-Al mare basalts, alkali basalts, KREEP
Luna-20	14.02.72	3°31′N 56°33′E	Highland site, Amegino crater	Ferroan anorthosite
Apollo-16	21.04.72	8°60′S 15°31′E	Highland site, Descartes crater	Ferroan anorthosite
Apollo-17	12.12.72	20°10′N 30°46′E	Mare Serenitatis	Low-Ti basalts
Luna-24	22.08.76	12°45′N 62°12′E	Mare Crisium	Very-Low-Ti mare basalts

impacts (just like shergottites from the surface of Mars). These data increase statistically the petrological scope of the sample collection, but since they have no exact geographical address, they are much less valuable than the samples collected directly on the Moon.

The following petrological descriptions of lunar rocks are based upon the following references: Lunar Sourcebook (Heiken et al. Eds., 1991); Luna – a step in technologies for exploration of the solar system, (in Russian) (Legostaev and Lopota, Eds., 2011).

3.5.1 Highland rocks

The rocks forming the continental areas of the Moon account for ∼ 95 % of the volume of the upper lunar crust.

They include a number of petrological series (Fig. 3.6).

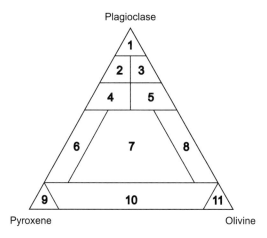

Figure 3.6. Classification scheme of lunar rocks (from Weizorek et al., 2006). 1 Anorthosite, 2 Noritic (Gabbroic) Anorthosite, 3 Troctolitic Anorthosite, 4 Anorthositic Notite (Gabbro), 5 Anorthositic Troctolite, 6 Norite (Gabbro), 7 Olivine Norite (Gabbro), 8 Troctolite, 9 Pyroxenite, 10 Peridolite, 11 Dunite.

The feldspar lunar crust is primarily composed of rocks with a high content of calcium plagioclase and iron oxide – ferroan anorthosites. The percentage of FeO varies from ∼ 1 % (pure anorthosites) to 15 % (Ferroan norites) with an average value of 4 %. Concentration of TiO_2 is low (∼ 0.5 % by weight). Thanks to the prevailing plagioclase, the rocks feature a high percentage of Al_2O_3 (> 24 %).

The average formation depth of these rocks does not exceed 25 km. Their formation is connected with surfacing of relatively light plagioclase crystals during crystallization and stratification of the hypothetical ancient magmatic ocean of the Moon. These were the first solidified rocks. Ferroan anorthosites are the oldest lunar rocks with an estimated age of 4.44–4.45 billion years. Samples of this suite prevail at the landing sites of the spacecrafts Apollo-16 and Luna-20.

Apart from anorthosites, continental rocks include magnesian magmatic rocks (the Mg-suite) (Fig. 3.7). The principal rocks in this suite are troctolite, spinel troctolite, norite, gabbronorite, anorthosite norites and ultramafic rocks. The main mineral of those rocks is plagioclase (mostly anorthite). Samples also include orthopyroxenes (enstatite and bronzite), clinopyroxenes (diopside, augite, pigeonite) and olivine.

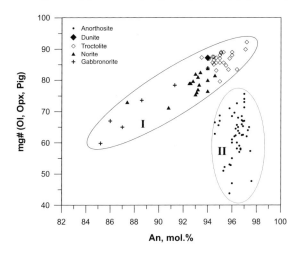

Figure 3.7. Magnesium factor mg# = Mg/(Mg + Fe) and anorthite content in plagioclase in the rocks of: I-magnesian suite and II-ferroan anorthosites (Ariskin, 2007).

The percentage of MgO varies from 45 % by weight in dunites to 7 % by weight in anorthosite norites; the percentage of Al_2O_3 ranges from < 2 % by weight in dunites to 29 % by weight in troctolite anorthosites. The percentage of TiO_2 does not exceed 0.5 % by weight. The age of the rocks is within 4.1–4.5 billion years. They are deemed to have emerged in the lower crust, at a depth of 30–50 km, and were implemented as intrusions into the ferroan anorthosite substrate and then uncovered during formation of large impact basins. There is portal overlap of the magnesian suite rock ages with those of the feroan anorthsites, which means it is likely they formed at approximately the same time.

A number of samples of the alkaline suite were collected at the landing places of Apollo-12, -14 and -15 (alkaline suite).

Typical minerals of the alkaline suite rocks are plagioclase, clinopyroxene, orthopyroxene, potassium feldspar, quartz, apatite, merrillite, ilmenite, chromspinel, fayalite, zircon, baddeleyite, troilite and the metallic phase of Fe-Ni. The rocks are relatively rich in potassium (0.3–0.5 % by weight) and sodium (1.25–1.6 %), considering that the Moon in general is poor in alkaline elements. The iron content varies from 0.4 % by weight in alkaline anorthosites up to 17 % by weight in alkaline norites; the TiO_2

content ranges from less than 0.5 % by weight up to 5 % by weight, and that of Th from 5 g/t up to 12 g/t, with a maximum content of 40 g/t. Quartz monzodiorites contain 65–75 % by weight of SiO_2, less than 10 % by weight of FeO, 3–8 % by weight of K_2O, and 1–2 % by weight of TiO_2. The same rocks feature high contents of REE, Zr, Hf, Rb, Cs, Nb, Ta, Th and U.

The age of the alkaline suite rocks is estimated at 3.8–4.3 billion years. Formation depth of the rocks is estimated at up to 2 km, i.e., the upper lunar crust.

Apart from the above principal petrological suites, there are some other rocks in the continental part of the Moon; these are mostly brecciated products of impact processing of the primary magmatic rocks as well as non-maria basalts and other rocks in low volumes.

3.5.2 Maria rocks

Maria rocks are relatively widely spread on the lunar surface (about 30 % of the visible hemisphere surface). However, volumetrically they are a minor component of the lunar crust. The average thickness of the basalt layer is about 400 m, increasing in some locations up to 1,000 m (Weiczorik et al., 2006). The total volume of extruded mare basalts is 10^7 km^3, i.e., only 0.1–0.5 % by the volume of the crust (Head and Wilson, 1992; Weiczorik et al., 2006).

The prevailing type of lunar maria rocks are maria basalts. The lunar maria basalts are subdivided into several groups with regard to their titanium, aluminum and potassium contents. Basalts with high titanium content ($TiO_2 > 8$ % by weight) were collected by the Apollo-11 and Apollo-17 missions. Basalts with low titanium and aluminum contents (TiO_2 2–6 % by weight, $Al_2O_3 < 12$ % by weight) are present in the samples brought back by the Apollo-12 and Apollo-15 missions. Basalts with low titanium content, rich in aluminum (TiO_2 3–6 % by weight, Al_2O_3 12–15 % by weight): this type includes basalts delivered by Luna-16. Basalts with very low titanium content ($TiO_2 < 1$ % by weight) were found in the sample delivered by Luna-24.

Further types identified are: low potassium, low titanium basalts with K_2O content of around 0.1 %, high potassium, high titanium basalts with K_2O content of around 0.3 %, extremely high potassium basalts with K content of 0.9 % by weight. As regards the mineral composition, differences between the above groups are represented by varying contents of Ti-bearing ilmenite and of feldspar, which account for most of the aluminum and alkaki (Longhi, 2006) .

The maria basalts are deemed to have originated from partial melting of the inner Moon at a depth of up to 400 km, and, therefore, their composition must to a certain extent represent the lunar mantle.

Vented basalt lavas filled the impact cavities of the maria less than 4 billion years ago, when intensive bombardment was over.

Within the maria formations, one comes across a pyroclastic material characterized by green and orange glass balls. They are deemed to have resulted from sprays of lava fountains. Such orange and green glass is found in the sample collections of Apollo-15 and Apollo-17.

A special type of rock is known as the KREEP basalts. The name is an acronym: Potassium (K), Rare Earth Elements (REE), Phosphorus (P). There are high Al_2O_3 (13–16wt%) and high FeO (9–15%) content basalts with anomalously high concentrations of incompatible trace-elements 100–150 times higher than chondritic concentrations (Weiczorik et al., 2006). Apart from the KREEP constituent, the difference to the maria basalts lies in the presence of iron olivine (fayalite) instead of magnisian olivine (forsterite) and in the high content of plagioclase. Pyroxene is represented by pigeonite and augite (clinopyroxenes).

Samples of the KREEP-basalts were collected by the Apollo-15 mission near the Apennine area, around Mare Imbrium and near Crater Aristillus.

3.6 Lunar chronology

Lunar chronology is based on two approaches: 1) absolute dating of samples brought to the Earth by the spacecrafts of the Apollo and Luna missions, and 2) statistics of impact craters covering parts of the lunar surface.

The larger the crater is, the slower it degrades. The average life of a 1 km crater is 5 million years, of a 100 km crater 250 Ma (million years), and of a 300 m crater 1.3 Ga (billion years) (Basilevsky, 1976).

The older a location is, the more craters it will contain per unit of area. By comparing the absolute age of the samples brought to the Earth from the landing places of the Apollo spacecrafts with the observed crater density in those areas, it has been possible to develop an age-measuring scale based on crater statistics.

As regards frequency and dimensions of the impact craters, the early history of the Moon is subdivided into the Pre-Nectarian period (older than 3.92 Ga), the Nectarian period (3.92–3.85 Ga) and the Early Imrbian period (3.85–3.8 Ga) (Wilhelms, 1987).

Analyses of lunar samples collected from different landing sites have allowed the dating of some major impact events and structures: Nectaris: 3.9 Ga, Crisium: 3.895 Ga, Serenitatis: 3.893 Ga, Imbrium: 3.85 Ga (Snyder et al., 2000). Crater-counting methods indicate that there are basalts of roughly 1 Ga (Heinsinger and Head, 2003): Autolycus: 2.1 Ga, Aristillus: 1.29 Ga, Copernicus: 0.8 Ga.

What is the age of the Moon? The answer to this question depends on our understanding of the origin of the Moon. The solar system started 4.568 Ga. The segregation of the Moon's core cannot have occurred earlier than 50 million years after the beginning of the solar system according to Hf–W data (Kliene et al., 2009).

If segregation of the lunar core occurred simultaneously with accumulation of the Moon then the Moon formed 4.51–4.52 Ga. If segregation of the lunar core occurred after the Moon was fully accumulated then the Moon could have formed earlier.

Table 3.2 lists ages of the most ancient lunar rock found so far. It is seen here that the ferroan anorthosites are the most ancient rocks to have appeared on the Moon surface.

Table 3.2. Age estimates for the oldest lunar rocks.

Object	Age		Isotope system	Reference
	Before present (Ga)	After solar system formation (Ma)		
Ferroanorthosite 60,025	4.5 + 0.01	67 ± 10	^{206}Pb/^{207}Pb	Hanan and Tilton, 1987
	4.45 ± 0.1	117 ± 100	Recalculated data of Hanan and Tilton (1987)	Halliday, 2008
Lunar crust	4.46 ± 0.04	107 ± 40		Norman et al., 2003
Descartes breccia 67,215	4.47 ± 0.02	97 ± 20	U–Pb	Tera et al., 1973
Ferroanorthosite 67,016	> 4.5(*)	< 70	^{147}Sm-^{143}Nd	Alibert et al., 1994
Ferroanorthosite 62,236	4.29 ± 0.03	277 ± 30	^{147}Sm-^{143}Nd	Borg et al., 1999
	4.45 ± 0.01		Sm-Nd	
KREEP	4.42 ± 0.07	137 ± 70		Nyquist and Shih, 1992
Norite from breccia 15,445	4.46 ± 0.07	107 ± 60	Sm-Nd	Shih et al., 1993
	4.44 ± 0.02	127 ± 20	Sm-Nd	Carlson and Lugmair, 1988
Ferroanorthosite 60,025	4.34	228	Sm-Nd	Carlson and Lugmair, 1981
Mg-suite	4.43	137	Rb-Sr	Nyquist et al., 1981
Mg-suite	4.46	107	U–Pb	Edmunson et al., 2008
Mg-suite				

*4.562 ±0.068

The geological history of the Moon started with the magma ocean (Warren, 1985). Its depth was about 500 km. As it cooled the magma ocean differentiated. When plagioclase solidified it floated, due to its lower density, and formed the early anarthositic crust of the Moon. The mafic cumulates are considered the source of the mare basalts. The residual liquid, enriched in large ion lithophilic elements (KREEP), could be remobilized by later magmatic cumulates to form basalts with a KREEP component (Snyder et al., 2000).

Mare basalt volcanism occurred during the great bombardment and formation of the large impact structures. The basalt lava that filled them up ranges in age from 3.6 to 3.9 Ga. The age of low titanium Al-bearing basalts at the landing site of Apollo-17 is estimated at $\sim$ 3.9 billion years, and that of the extremely low titanium basalts at the Luna-24 landing site at around 3.3 billion years. The age of high titanium basalts at the Apollo-11 site is estimated within the range of 3.5–3.8 billion years, of high potassium, high titanium basalts at the same place is about 3.55 billion years, and that of low titanium basalts near the Apollo-12 and Apollo-15 sites is in the range of 3.08–3.37 billion years.

The average KREEP model age is 4.42 ± 0.07 Ga (Nyquist and Shih, 1992). The ages of ferroan anorthosites and Mg-suite rocks overlap, suggesting that by about 4.4 Ga both types of magmas were being produced on the Moon (Snyder et al., 2000).

The Moon differentiation was completed by 4.49 Ga ($\sim$ 80 million years after the solar system formed) assuming a chondritic uniform reservoir bulk composition for the Moon (Edmunson et al., 2008).

Along with the most ancient ferroan anorthosites there are relatively young ones. For example, sample 62,236 (Apollo-16) yields an age of 4.29 ± 0.03 Ga (Borg et al., 1999). This is inconsistent with the concept of a short-living magma ocean.

Total solidification of the lunar magma ocean with a depth of $\sim$ 500–1,000 km requires only 10 million years, and solidification of the first 80 vol% of the magma ocean to the point that plagioclase begins to float requires less than 1000 years (Elkins-Tanton et al., 2011). However, the range of ages of ferroan anorthosites is almost 200 million years. Hence lunar crust magmatism and lunar crust formation continued during this time. A possible explanation is tidal heating. It has been demonstrated that tidal heating due to interaction with the early Earth is sufficient to continue magma ocean solidification for 200 Ma (Meyer et al., 2010). Early dynamical heating could have been comparable to internal radiogenic heating (Weiczorik et al., 2006).

Tidal energy in the early period of lunar history might have been able to promote convection and dynamo if the core was metallic (Weiczorik et al., 2006).

Formation of a layer of denser phases including ferrous silicates and ilmenite at the final stage of magmosphere crystallization may have caused gravitational instability and overturn that led to the influx of primitive magnesian magmas into the upper horizons (Hess and Parmentier, 1995; Elkins-Tanton et al., 2002).

3.7 Internal structure and temperature

The internal lunar structure is reconstructed via studies of the elastic (seismic) properties of the Moon, of the gravity field and of the magnetic field as well as via analysis of composition of the rocks collected on the surface and representing the composition of the upper lunar layer.

Contemporary knowledge about the internal lunar structure is based upon the data obtained from the four seismic stations where the Apollo spacecraft recorded seismic behavior of the Moon for 8 years from July 1969 till September 1977.

The tectonic activity of the Moon is incomparably lower than that of the Earth. The annual output of seismic energy on the Moon is around 10^{10} J, compared to around 10^{18} J on the Earth (Goins et al., 1981).

Most seismic events take place within the depth range from 50 to 220 km (Khan et al., 2000). Another active interval is deep focus moon-quakes in the 950–1,000 km depth range.

A clear stratification is observed on the Moon. However, limits of the strata are established in different ways by different scholars, depending on their approach to seismogram interpretation. The estimated thickness of the crust ranges from 30 to 70 km (on average 45 km). The lower limit of the upper mantle can be quite positively established at a depth of 500 km. Sometimes a transition area (the middle mantle) is identified between the upper mantle and the lower mantle at a depth of 500–650 or 500–750 km. No seismic data from Apollo are available regarding areas below 1,000–1,100 km.

Only for the upper part (0–300 km) of the lunar lithosphere do scholars more or less concur in estimations of the velocity of elastic waves: average value $V_S = 4.48 \pm 0.05$ km/s and $V_P = 7.71 \pm 0.06$ km/s according to the data of Goins et al., (1981); Nakamura (1983); Lognonne et al., (2003); Khan and Mosegaard (2001). Estimates by the same authors for the lower mantle vary greatly from each other. As an example, Fig. 3.8 shows estimates made by Nakamura (1983) and Lognonne et al., (2003).

The first estimates on thickness of the lunar crust were close to about 70 km on average. The seismic data were later reanalyzed toward lower values: 45 ± 5 km (Khan et al., 2000), 38 ± 8 km (Khan and Mosegaard, 2001), 30 ± 2.5 km (Lognonne et al., 2003). Weiczorik et al., (2006) estimated the thickness of the lunar crust at 49 ± 1.6 km. The maximum crustal thickness is 85 km (located at 199° E, 4° N), and the minimum thickness is about zero, located near the edge of South Pole Aitken beneath Mare

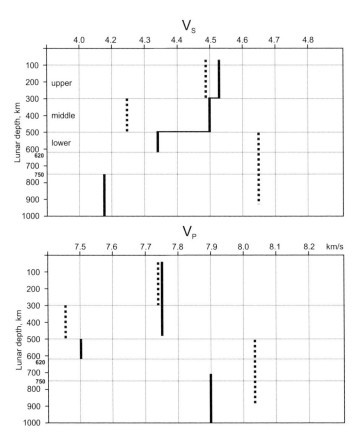

Figure 3.8. Lunar seismic P-and-S-wave velocities (straight line – author Nakamura, 1983, dotted line – author Lognnone et al., 2003).

Moscoviense (151° W, 36° S), which would have been formed by strong impact and resulting mantle uplift. Fei et al. (2012) recently suggested that the most likely average lunar crustal thickness is 42.2 km. We accept here 45 km for the average thickness of the lunar crust.

The crust is stratified to upper anorthositic and lower noritic layers of approximately equal thickness $\sim$ 25 km on average. The volume of the crust comprises 8–10 % of the total lunar volume. Its average density is 2.9 g/cm^3.

There are no samples directly representing the lunar mantle. The composition of the upper mantle can be inferred through study of mare basalts and volcanic glasses, which are derived from the upper mantle. Its composition and structure can also be verified by analysis of the seismic data.

The sharp increase of velocities deeper than 500 km is interpreted as evidence of change in the chemical (mineralogical) composition of the mantle from a differentiated upper part to a primordial lower mantle (Nakamura, 1983; Hood, 1986).

Unlike the Nakamura model, P-wave velocity in the Khan et al. (2000) model is essentially constant or increases a little with depth (8.1 ± 1.5 km/s). Below 500 km the velocity significantly increases to as much as 10.0 ± 1.5 km/s, and at a depth of about 620 km it returns to the value of 8.1 km/s. Then it again gradually increases with depth to 11.0 ± 1.2 km/s at 950 km. Thus the maximum P-wave velocity occurs in the interval of deepest lunar moon-quakes.

Elastic properties depend both on the rock composition and on the temperature.

The upper mantle is characterized by a high seismic inverse dissipation factor Q (about 4,000–7,000), which means an anomalously long continuation of seismic signal (Lognonne, 2005). The high Q-values are indicative of hard rock, low porosity and lack of volatiles. In the upper mantle the higher temperature limit can be inferred from the mascon anisostasy (Lambeck and Pullan, 1980). It requires a temperature less than 800 °C, to keep mascons stable during almost 4 Ga. In the interval 500–600 km the Q-factor decreases to a value near 1,500 (Toksöz et al., 1974). However, this is still indicative of temperatures well below the solidus. Location of most deep moon-quakes at a depth of 850–950 km still requires quite solid rock to allow stress accumulation (Hood and Zuber, 2000). Finally at depths greater than 1,000 km S-wave velocity decreases and Q-values decrease to $\sim$ 100, implying temperatures approaching the solidus. L. Hood and M. Zuder (2000) suggest the following temperature profile: $\sim$ 750 °C at 300 km depth, $\sim$ 1,200 °C at 800 km depth, and $\sim$ 1,400 °C at 1,100 km depth.

In the upper crust, the average percentage of plagioclases is 82 %, which corresponds to 27–29 % of Al_2O_3. The percentage of plagioclases in the lower crust is 71–75 %, and the content of Al_2O_3 is 18–20 %, respectively (Tompkins and Pieters, 1999).

It has been demonstrated that the pyrolitic composition is definitely not suitable for describing the chemical composition of the upper mantle, since it predicted unrealistically high temperatures at the respective depths. Satisfactory results are obtained for a pyroxenite composition with low Al and Ca contents: $\sim$ 400–500 °C at 100 km, 600–750 °C at 300 km, 750–1,000 °C at 500 km (Kuskov et al., 2002). In the lower mantle, reasonable temperature estimates of 1,250–1,350 °C are obtained for an association of olivine + clinopyroxene + garnet (i.e., a composition rich in refractory oxides: 4–6 % by weight CaO and Al_2O_3).

The average density of the mantle is estimated to be 3.36 g/cm^3. Unfortunately the Apollo seismic network could not determine the deep interior of the Moon and the size of the core.

As noted above, judging by the value of angular momentum, the Moon may have a small metallic or sulfide core weighing at maximum 5 % of its total weight. Based on studies of gravimetrical, magnetic and seismic properties of the Moon as well as on the results of the Lunar Laser Ranging (LLR), we are now able to make certain preliminary conclusions regarding existence and size of the core.

These studies have resulted in the conclusion that the Moon has a core consisting of a solid internal core with a radius ranging from 160 km or more (Weiczorec et al., 2006) to 240 km (Weber et al., 2011) and a liquid outer core with a radius of $\sim$ 330–350 km. The core may be surrounded by a layer of molten silicates from the lower mantle with a radius of 480 km (Weber et al., 2011) to 587 km (Weiczorek et al., 2006). The plentiful scientific harvest reaped by the LLR contained among other things the frequency dependence of the tidal dissipation in the Moon at monthly frequencies (Williams and Dickey, 2003). The slope of this dependence looks "improper", in that the tidal quality factor slowly decreases with the increase of the frequency (Williams and Boggs, 2009), a behaviour incompatible with what we might expect from our knowledge of seismology (Karato and Spetzler, 1990). Comparing the measured frequency dependence with a theoretical model based on realistic rheology, Efroimsky (2012) demonstrated that the "improper" slope registered by the LLR can be reconciled with the theory only with a low average viscosity of the Moon. Were the Moon a uniform viscoelastic body, it would have a surprisingly low viscosity of $3 \cdot 10^{16}$ Pa·s. This estimate indicates that the lunar lower mantle is likely to contain a high percentage of partial melt. The pressure at the core-mantle boundary is about 4 GPa. The temperature which might be expected at the core-mantle boundary is unlikely to exceed 1,500 °C. As the melting temperature of pure iron is 1,690 °C for the pressure $\sim$ 4 GPa, and the LLR predicts partly liquid core, this means that the lunar core is probably not a pure iron core.

The presence of sulphur decreases the melting temperature of Fe-FeS system to about 950 °C. Therefore if sulfur were present in the core, some portion of the core should be molten (Weiczorek et al., 2006).

Kuskov and Kronrod (1998) have estimated the radius of the core to be 310–320 km for a pure iron composition and 430–440 km for Fe-FeS eutectic composition. The size of the core would be even larger if the core were assumed to be composed instead of a less dense titanium and iron-rich silicate material (Weiczorik et al., 2006).

It should be recognized that precise composition, formation conditions and thermal evolution of the lunar core are not well constrained at present.

Chapter 4

Similarity and difference in composition of Earth and Moon

4.1 Iron content

The fact that the Moon contains less iron than the Earth, and that it probably has no iron core at all, was noted long ago due to the sharp difference in their densities, which (under normal pressure) are 4.45 and 3.3 g/cm^3 for the Earth and the Moon, respectively. The Earth has a massive core consisting of iron and nickel, plus up to 10% of lighter elements, according to the geophysical data. H, C, O, Si and S have been considered as possible light elements.

Total iron content of the Earth is around 32.5%, whereas total iron content of the Moon is estimated at less than 15% (Fig. 4.1).

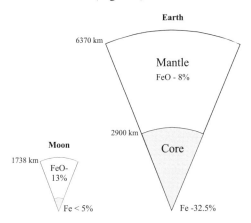

Figure 4.1. Comparative iron content in the Earth and the Moon.

Although the Moon, as a planetary body, contains much less iron than the Earth, the iron content in its mantle is proportionally higher than that in the Earth's mantle. The average concentration of FeO in the Earth's mantle is estimated as 8%. The estimates for the Moon range from 13 to 18% (Delano, 1986; Taylor, 1986). The lower limit is probably a more plausible estimate (Jones and Delano, 1984). However, it is still significantly higher than the value accepted for the Earth's mantle.

Therefore, the difference between the Earth and the Moon lies not only in the fact that the Earth as a whole contains much more iron, but also in the fact that concentration of iron in the silicate part of the Moon is higher than that in the silicate part of the Earth.

4.2 Redox state

Unlike the Earth's mantle the lunar mantle does not contain any traces of Fe_2O_3. The polyvalent elements like Cr and Ti always are present in their lower valence (Haggerty et al., 1970). This is corroborated with mineralogical observations of Fe-rich olivine, cristobalite, and metal in the mesostasis of mare basalts (Wieczorek et al., 2006). Presence of olivine + spinel + metal inclusion in orange glass beads suggests that metal-melt equilibria and the Cr content of the olivine and spinel indicate a pre-eruptive f_{O_2} of 1.3 log units below IW (Weitz et al., 1997).

Many other direct and indirect lines of evidence support a reduced mantle with f_{O_2} at or below the IW buffer (Weiczorek et al., 2006) in contrast to the oxidated state (QFM buffer) of the present Earth's upper mantle.

Thus, there exists a considerable difference between redox conditions in the mantles of the Moon and the Earth. However, this difference may prove to be secondary. The Earth has a huge metallic core. If we were to balance the silicate part of the Earth and its core, the redox condition of the Earth in total would be as reduced as that of the Moon (Galimov, 1998). It is quite possible that immediately after accumulation, the atmosphere on the early Earth was strictly reduced. Evidence of this can be found, for example, in the fact that the ratio K/Na > 1 required for the genesis of protein life in a hydrosphere occurs in the case of $CH_4/CO_2 > 1$ in the early atmosphere (Galimov and Ryzhenko, 2008; Galimov et al., 2011).

Probable evolution of the redox condition in the Earth's mantle has been studied in a number of works: (Kasting et al., 1993; Stevenson, 1983; Lecuyer and Ricard, 1999; Frost et al., 2004; Galimov, 2005)

In particular, we have shown (Galimov, 1998; Galimov, 2005) that slow build-up of the core within geological periods can at the same time explain: (1) the existence of a heat source maintaining a superdiabatic temperature gradient (and related convection) in the mantle of the Earth, and (2) oxygen influx into the mantle due to disproportioning of valency of iron: $3Fe^{2+}_{(descending\ convective\ branch)} = Fe^{0}_{(core)} + 2Fe^{3+}_{(ascending\ convective\ branch)}$ (Fig. 4.2).

If this is the case, there is no difference between the initial redox conditions of the Moon and the Earth. As regards the difference observed, it took place because the process of convection on the Earth led to gradual oxidation of the mantle until it reached its current condition of the QFM buffer (this condition had been reached by 4 Ga already). However, there was no such mechanism on the Moon. That is why it has preserved its original redox state. This is consistent with the idea that the lower mantle of the Moon is composed of "primitive" unmelted materials (Wieczorek et al., 2006).

Hence, it follows that the observed **difference in redox characteristics between the Moon and the Earth cannot be used as a constraint on the Moon origin.**

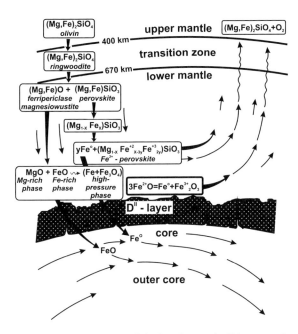

Figure 4.2. Model of the redox evolution of the Earth mantle. Disproportionation of iron provides growth of the core through accommodation of Fe and FeO from the mantle to the core. Excess of the oxygen was entrained with the backward convection flow with some admixture of corederived material. The DU-layer is supposed to be a site where core and mantle materials interpenetrate, interfuse and interact.

4.3 Volatiles

The problem of iron deficit in the Moon has traditionally played a pivotal role in the discussion of the Moon origin. Another fundamental difference between the Earth and the Moon is the depletion of the Moon in volatile elements. It is known that all cosmic bodies are depleted in volatiles relative to their solar abundance. This is true for carbonaceous chondrites, although CI carbonaceous chondrite material is considered to be the most primitive material. Ordinary chondrites in turn, are depleted in volatiles relative to carbonaceous chondrites. The composition of the Earth shows a deficit in volatiles compared with the ordinary and carbonaceous chondrites. The Moon is depleted in volatiles to a greater extent and is similar in this respect to achondrites, which are considered fragments of the silicate shell of differentiated asteroids (e.g., Vesta). Thus, on the one hand, the Moon is strongly depleted in volatile components compared with the Earth, and this fact must be explained by the mechanism of the formation of the Earth-Moon system. On the other hand, the volatile-poor composition of the Moon reflects a general phenomenon characteristic of matter near the Sun, at least in the inner (to Jupiter) zone of the solar system.

The concentrations of volatiles in meteorites, the Moon, and the Earth were estimated by various authors (Palme et al., 1998; Humayun and Cassen, 2000). Figure 4.3 shows the diagram of volatile depletion on the Moon according to Ringwood (1986) with some modifications (Galimov, 2004).

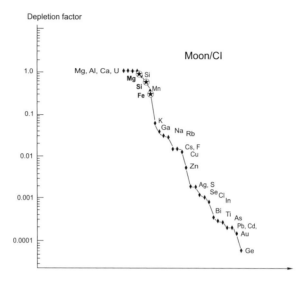

Figure 4.3. Depletion of the Moon according to Ringwood (Ringwood, 1986) with modifications for Si, Mg, and Fe (Galimov, 2004).

An important feature of the volatile depletion on the Moon is that the degree of depletion is controlled by the volatility of the element and is practically independent of its atomic mass. The Moon is significantly depleted in K and Na, which are also low on the Earth. The Moon is even more depleted in Rb. Rubidium is heavier than K but more volatile. The K/U ratio is 60×10^3 for carbonaceous chondrites, 11×10^3 for the Earth, and 2.5×10^3 for the Moon. The Rb/Sr ratio is 0.3 for carbonaceous chondrites, 0.031 for the Earth, and 0.009 for the Moon. The Moon is extremely depleted in lead, which is one of the most volatile heavy elements. It could be gravitationally retained not only by the Moon but even by smaller bodies. Consequently, the process of evaporation must have occurred from relatively small bodies and/or through hydrodynamic escape.

Another important characteristic is that the process responsible for the volatile loss occurred very early in the lunar history. Lunar samples have very low $^{87}Sr/^{86}Sr$ ratios in comparison to the Earth's rocks. This means that the Moon lost ^{87}Rb, volatile radioactive precursor of ^{87}Sr, very early.

The Moon is characterized by anomalously high ratios of the radiogenic lead isotopes ^{206}Pb, ^{207}Pb, and ^{208}Pb to non-radiogenic ^{204}Pb. For example, lunar rocks are characterized by $^{206}Pb/^{204}Pb$ ratios varying in the range 160–190, while for the Earth the ratios around 18–20 are typical (Edmunson et al., 2008). This also suggests that the catastrophic loss of lead took place very early, when it was represented mainly by the primordial composition including ^{204}Pb, and the subsequent accumulation of lead throughout lunar history occurred at the cost of the decay of radioactive isotopes of the refractory elements U and Th.

The third characteristic of volatile loss from the Moon is that this process has left no isotope fractionation record. It is known that evaporation is accompanied by isotope fractionation. Wang et al. (1994) showed that evaporation of 40 % melt changes the (^{30}Si/^{28}Si) ratio of the residue by 8–10‰, and (^{26}Mg/^{24}Mg), by 11–13‰, in accordance with the simple Raleigh distillation equation. Given the observed degree of potassium depletion on the Moon, the isotopic composition of potassium (^{41}K/^{39}K) should have changed by more than 90‰ (Humayun and Clayton, 1995). In fact, no isotopic shifts were reported for these elements in various classes of cosmic bodies, including the Moon.

In Chapter 7 we will discuss this phenomenon in more detail in relation to the considered hypotheses of the Moon origin.

4.4 Refractory elements

Refractory elements have very low volatility. In vaporization processes they are retained in a condensed phase to the highest temperature point. Such elements comprise Al, Ca, Ti, Sr, Sc, Y, REE, Zr, Hf, Th, V, Nb, Ta, Mo, W, U, Re, Os, Rh, Ir, and Pt. These elements are geochemically very different and may show pronounced lithophile or siderophile properties. They are significantly fractionated by magmatic processes. It is important, however, that they behave similarly in high-temperature cosmogonic processes relating to the evaporation and condensation of materials. As a result, although they may be concentrated in different planetary shells, the proportions of the refractory elements remain constant in the planetary body as a whole (Table 4.1).

It is noteworthy that this group includes the rock-forming elements Al, Ca, and Ti, as well as U and Th, which are responsible for radioactive heat generation.

Table 4.1. Concentrations of some refractory elements in the Earth, Moon, and carbonaceous chondrites.

Element	CI	Earth	Moon
Al_2O_3 (%)	2.44 (1)	3.64 (1)	6.0 (1)
CaO (%)	1.89 (0.77)	2.89 (0.79)	4.5 (0.75)
TiO_2 (%)	0.11 (0.045)	0.16 (0.04)	0.3 (0.05)
U (ppb)	8 (3.27×10^{-7})	20 (5.5×10^{-7})	33 (5.5×10^{-7})
Sr (ppm)	11.9 (4.9×10^{-4})	17.8 (4.9×10^{-4})	30 (5.5×10^{-4})
La (ppb)	367 (1.5×10^{-5})	551 (1.5×10^{-5})	900 (1.5×10^{-5})

Note: Numerals in parentheses are concentrations normalized to Al_2O_3 content in particular object.

There are different estimates for the chemical composition of the Moon including the content of refractory elements (Table 4.2). Some authors believed that the concentrations of Al, Ca, Ti, and other refractory elements in the Moon are identical or only slightly different from the terrestrial values (Ringwood, 1986; Wanke and G. Dreibus, 1986).

Table 4.2. Estimates of various authors for the chemical composition of the Moon.

Component	(Buck and Toksoz, 1980)	(Warren, 1986)	(Taylor, 1986)	(Ringwood, 1986)	(Jones, Delano, 1984)	(O'Neil 1991)
SiO_2	48.4	46.0	43.5	43.2	43.5	44.6
TiO_2	0.40	0.3	0.3	0.3	0.29	0.17
Al_2O_3	5.00	7.00	6.0	3.7	5.8	3.9
FeO	12.9	12.4	13.0	12.2	16.1	12.4
MgO	29.0	27.6	32.0	36.9	29.9	35.1
CaO	3.83	5.5	4.5	3.0	4.6	3.3
Na_2O	0.5	0.6	0.09	0.06	–	0.05
K_2O	–	0.06	0.01	–	–	–

However, recent studies support the idea that the Moon is significantly enriched in refractory elements relative to the Earth.

The lunar crust is about 50 km thick and enriched in refractory elements with an estimated Al_2O_3 content of 28–30 %. The underlying upper mantle is 500 km thick and is depleted in Al_2O_3 (Fig. 4.4).

Composition of the upper mantle can be inferred from a study of mare basalts, which are the product of partial melting of the upper mantle. The Al_2O_3 content estimated on this basis is less than 1 % (Hood and Zuber, 2000).

The middle and lower mantle (> 500 km depth) is inaccessible to sampling and can be characterized only by geophysical methods. Based on the seismic model by Nakamura (Nakamura, 1983), the elastic properties of the lower mantle suggest the presence of 22–28 % garnet, which is equivalent to an Al_2O_3 content of 7–10 % (Hood, 1986). Kuskov and coauthors (Kuskov and Fabrichnaya, 1994; Kuskov and V. A. Kronrod, 1998) demonstrated that the velocity profile by Nakamura agrees with their synthetic curve for a depth range of 500–1,000 km at an Al_2O_3 content between 6 and 7 %. Thus the concentration of Al_2O_3 in the bulk Moon accounting for its distribution in particular shells is consistent with values between 5.3 and 6.9 %, which is close to Taylor's estimate (Table 4.2). In Table 4.3 the lunar composition according to the data of S. Taylor is compared with the compositions of the silicate part of the Earth and with the CI chondrites (CI).

In the following chapters we will show that the high content of refractory elements on the Moon is of principal importance for selection of a lunar genesis model.

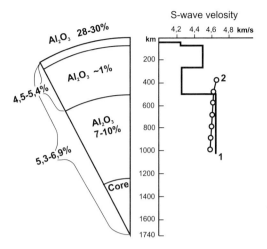

Figure 4.4. Proposed internal structure of the Moon: 1) seismic profile based on the published results of the Apollo mission (Nakamura, 1983); 2) synthetic seismic profile for a rock containing 6 % Al_2O_3 (Kuskov and Fabrichnaya, 1994).

Table 4.3. Comparison of the compositions of the Earth, Moon, and CI chondrites.

Component	CI	Silicate Earth	Moon
SiO_2	34.2	49.9	43.4
TiO_2	0.11	0.16	0.3
Al_2O_3	2.44	3.64	6.0
FeO	35.8	8.0	13.0
MgO	23.7	35.1	32.0
CaO	1.89	2.89	4.5
Na_2O	0.98	0.34	0.09
K_2O	0.10	0.02	0.01

4.5 Thermal evolution

The thermal structure of the Moon has been discussed above (Section 3.7). A comparison of the thermal state and evolution of the Earth and the Moon reveals a fundamental difference.

In the Earth the value of the global Earth heat flow $(4.0$ to $4.1) \times 10^{13}$ W (Verhoogen, 1980; Davis and Richards, 1992) significantly exceeds the amount of heat provided by radioactive decay (2.4×10^{13} W). The contribution of all other sources (tidal dissipation, crystallization of the inner core etc.) do not exceed 0.5×10^{13} W. It has been shown (Galimov, 1998, 2005) that this disbalance can be eliminated by the process of low-rate growth of the Earth's core during geological time. This process is suggested to occur at the expense of extraction of FeO by the core from the mantle directly or by Fe-disproportionation ($3Fe^{2+} = F^0 + 2Fe^{3+}$) in the course of global convection. This process could provide heat emission comparable with radioactive heat, and it maintains the superadiobatic temperature gradient necessary for convection. Simultaneously it explains redox evolution of the Earth's upper mantle from reduced to oxidation state (see above).

In the case of the Moon the situation is different. Putative pristine composition of the lunar middle and lower mantle excludes long-term convection. Therefore the amount of heat produced should be confined by radioactive heat. This actually occurs.

In the Earth the contributions to the present heat flow from the main heat producing elements are approximately equal: Th $- 8.5 \times 10^{12}$ W, U $- 8.0 \times 10^{12}$ W, K $- 7.5 \times 10^{12}$ W (Galimov, 2005).

The Moon is enriched in refractory elements including U and Th: U $-$ 33 ppb compared to 20 ppb in the Earth. On the other hand the Moon is depleted in volatile K. The K/U ratios are $11 \cdot 10^3$ and $2.5 \cdot 10^3$ for the Earth and the Moon respectively. It follows that the Moon is depleted in K about 2.6 times relative to the Earth. Summing of the U, Th and K components of lunar radioactive heat yields 51.6×10^{10} W for the present lunar heat flow.

The present specific heat flow through the lunar surface is estimated to be $11 \div 18$ (between 11 and 18) mW/m^2 (Langseth et al., 1976; Hood, 1986) inferred from direct measurements at the Apollo-15 and Apollo-17 landing sites. The total heat flow is $(11 \div 18) \cdot 10^{-3} \times S_{\text{Moon's surface}} = (11 \div 18) \times 10^{-3} \times 3.8 \cdot 10^{13} = (41.8 \div 68.4) \cdot 10^{10}$ W.

Thus the calculated radioactive heat production in the Moon is within the limits of the range of the direct measurements. That means that there is approximately a **steady-state balance** between radioactive heat production and its loss through the lunar surface. No other significant heating sources are involved. And this is a **principal difference between present thermal states of the Earth and the Moon**.

4.6 Isotopic composition similarity

While the different iron contents of the Moon and the Earth have been regarded as the main composition difference between them and served as a template for testing this or that lunar genesis hypothesis, the main composition similarity between the Earth and the Moon lies in their indistinguishable isotopic compositions.

4.6.1 Oxygen

The most typical example is the coincidence between the isotopic characteristics of oxygen on the Moon and the Earth. Variations between ratios of $^{17}O/^{16}O$ and $^{18}O/^{16}O$ normally measured by means of $\delta^{17}O$ and $\delta^{18}O$ values exist in the rocks of both the Earth and the Moon. But their isotopic fractionation lines coincide completely.

Isotopic compositions of elements may change due to certain physical and chemical processes. There are certain isotopic effects that lead to re-allocation of isotopes between interacting chemical components: thermodynamic and kinetic isotopic effects. Fractionation of isotopes also takes place due to different degrees of mobility of lighter and heavier isotopes, for example, in cases of diffusion, evaporation, etc.

In all those cases we observe the phenomenon known as mass dependent isotopic fractioning. It means that when the mass difference between isotopes is, say, 2 atomic units – as, for example, in the case of ^{16}O and ^{18}O – isotopic fractioning will be around 2 times higher than when mass difference between isotopes is 1 atomic unit – as, for example, in the case of ^{16}O and ^{17}O. Variations of isotopic compositions are mass dependent for the overwhelming majority of natural processes. However, this rule is broken in some cases. First of all, this is linked with radiogenic isotopes. When the parent isotope decays, there emerges just one isotope from the pleiad of a given element. This results in non-proportional change of isotopic relations.

There are also certain chemical processes leading to non-proportional changes of isotopic composition – for example, radical reactions, kinetic behavior of which depends on super-fine interaction between electronic and nuclear spins (Buchachenko and Galimov 1976; Galimov, 1979). In this case, behavior of isotopes will be different depending on the nuclear spin of the isotope (for example, spin of the ^{13}C nucleus is ½, while spin of the ^{12}C is zero). Non-proportional fractionation of isotopes may accompany certain photochemical reactions because excitation frequency is different for various isotopic forms of the specific compound (Thimens and Heidenreich, 1983).

Mass independent isotopic fractioning is easy to register if the element has a number of isotopes, such as, for example, oxygen: ^{16}O, ^{17}O and ^{18}O.

Isotopic composition of oxygen from Earth samples varies within a wide range. But if one consider the three-isotope diagram $\delta^{17}O - \delta^{18}O$, values determining the isotopic composition of oxygen from the Earth samples are placed along one straight line that is running at an angle of ½ within the same coordinates. Meanwhile, there are a lot of objects in the solar system that lie outside the line of the Earth isotopic fractionation (Fig. 4.5).

The earliest explanations of the anomalous mass-independent oxygen isotope distribution favored a nucleosynthetic mechanism (Clayton, 1993). Then quantum symmetry driven isotopic fractionation for the gas-phase production of O_3 was suggested (Thimens and Heidenreich, 1983). Later photo-chemical self-shielding of CO (the most abundant O-bearing molecule in the solar nebula) has been widely adopted as the favored explanation (Clayton, 2002; Yurimoto and Kuramoto, 2004; Lyons and Young, 2005).

The CO-self-shielding mechanism suggests that photodissociated CO-molecules are enriched in ^{17}O and ^{18}O atoms, which react to produce compounds incorporated to dust, while $C^{16}O$ is assumed to remain in the gas phase.

Recently low temperature formation of ozone on cold ($T \sim 30\,\text{K}$) surfaces of interstellar dust has been thought to play a role in the mass-independent oxygen isotope pattern of solar system material (Dominguez et al., 2012). Some other reactions have been reported to produce a mass-independent oxygen isotope effect.

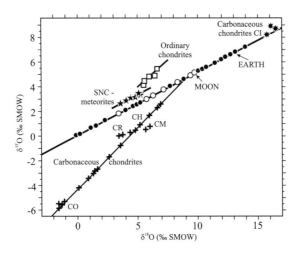

Figure 4.5. Oxygen isotope composition of different solar system bodies (R. Clayton's diagram).

In any event, the observed heterogeneity is a result of mixing of oxygen in different proportions from isotopically different reservoirs in the solar environment.

Recently data on the oxygen isotope heterogeneity in the early solar nebula have been supplemented by measurements of the grains from Comet 18P/Wild-Z collected by NASA's Stardust missions (McKeegan et al., 2006, 2011; Snead et al., 2012). Measurements of the oxygen isotope composition of particles from the comet tail yielded values scattered throughout the $\delta^{18}O$–$\delta^{17}O$ diagram (Fig. 4.6).

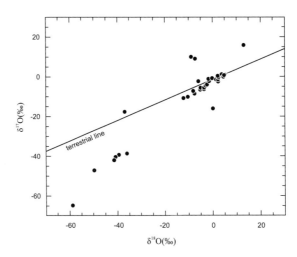

Figure 4.6. Oxygen isotope data of Wild-2 stardust materials (generalized for the different tracks, after Snead et al., 2012).

Section 4.6 Isotopic composition similarity

Against the background of significant oxygen isotope heterogeneity in the solar system the Earth and the Moon reveal a striking similarity. Their oxygen isotope composition values lie on the same mass-dependent isotope fractionation line. This **similarity is an important feature of the Earth-Moon system and a strong constraint on the Moon origin.**

4.6.2 Silicon

Silicon, just like oxygen, has three isotopes – ^{30}Si, ^{29}Si and ^{28}Si. However, no matter how the isotopic composition of silicon might change, the following rule is observed: $\delta^{29}\text{Si} = 0.52\delta^{30}\text{Si}$. In other words, isotopic fractioning of silicon is not affected by any mass-independent isotope effects.

Isotopic compositions of silicon in the Earth and lunar samples are close (Epstein and Taylor, 1970). The lunar mare basalts show δ^{30}Si-values from -0.35 to $-0.27\permil$ with an average of $-0.31\permil$, and the terrestrial oceanic basalts range from $-0.51\permil$ to $-0.31\permil$ averaging at $-0.38\permil$ (Georg et al., 2007).

The isotopic composition of Si in meteorites, including carbonaceous chondrite, ordinary chondrites and achondrites, is slightly, but distinguishably, different: δ^{30}Si from $-0.70\permil$ to $-50\permil$ with an average of $-0.58\permil$ (Georg et al., 2007).

Whatever was the cause of the difference between meteorites and the Earth-Moon material, the important issue is that **the Earth and the Moon have an identical silicone isotope signature**.

4.6.3 Titanium

Titanium is a refractory element comprising 5 isotopes: ^{46}Ti, ^{47}Ti, ^{48}Ti, ^{49}Ti, ^{50}Ti. Titanium isotopic compositions are reported in the ε notation, where ε is the deviation in part per 10^4 of the Ti-isotope ratio measured in the studied sample from the Ti-isotope ratio in a standard (terrestrial) sample. ^{47}Ti serves as the reference isotope, that is, the $\varepsilon\,^{50}$Ti-values are variations of ^{50}Ti/^{47}Ti ratios, and $\varepsilon\,^{46}$Ti-values are variations of ^{46}Ti/^{47}Ti ratios.

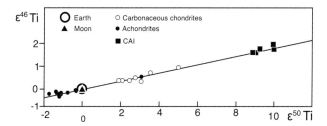

Figure 4.7. ε^{46}Ti – ε^{50}Ti – diagram for some solar system objects (data from Trinquir et al., 2009). The isotope variations are presented by a single line with slope 5.47 ± 0.27.

As seen from Fig. 4.7 there exist significant cosmochemical variations of Ti isotope compositions. The range of variations among different types of meteorites is about 7 units in the ε^{50}Ti scale (Trinquier et al., 2009).

At the same time a high precision Ti-isotope study on lunar samples has shown that **the Moon and the Earth are indistinguishable in their Ti isotope composition** (Zhang et al., 2011). The lunar samples have the same ε^{50}Ti, ε^{46}Ti, ε^{48}Ti as the Earth's sample within $\sim 0.1\,\varepsilon$ units.

4.6.4 Cromium

In contrast to oxygen, silicon and titanium, which have no radiogenic isotope, the Cr-isotope composition is subject to contribution from a parent radioactive nuclide ^{53}Mn.

Presence of decay products of short-lived radionuclides in many solar system objects indicates that the newly generated elements were injected into the solar nebula just before or during the solar system formation.

Isotope ^{53}Cr is a radiogenic daughter of ^{53}Mn, which decays with $T_{1/2} = 3.7$ Ma. The isotopic variations of the ^{53}Cr/^{52}Cr ratio are measured as the deviation in parts per 10^4 (ε-unit) from the terrestrial ^{53}Cr/^{52}Cr.

The Moon and Earth have indistinguishable ^{53}Cr/^{52}Cr ratios. For the Moon ε^{53}Cr ~ 0 (Carlson and Lugmair, 2000; Shukolyukov and Lugmair, 2000). It means that their parent material received the same portions of ^{53}Mn. At the same time the other solar system objects have different ε^{53}Cr values. The CNC-meteorites have on average ε^{53}Cr $= 0.24$, ordinary chondrites also uniform and have some higher ^{53}Cr excess: ε^{53}Cr ~ 0.48. The achondrites (howardites, eucrites, diogenites) reveal a range of ^{53}Cr excess from $\sim 0.39\varepsilon$ to 1.29 with average ε^{53}Cr ~ 0.57. The enstatite meteorite showed ε^{53}Cr ~ 0.17. Thus there exist small but quite clear differences in the relative abundance of the ^{53}Cr isotope in different solar system bodies. Shukolyukov and Ligmair (2000) have shown that the differences reflect ^{53}Mn heterogeneity, which correlates with heliocentric distance.

Thus the **coincidence of Cr-isotopic composition of the Earth and the Moon indicates that they originated from a common source material**.

4.6.5 Tangsten

Just like the ^{53}Cr isotope, ^{182}W has a short-living radioactive predecessor, ^{182}Hf, with a half-life of 9 million years, which turns into ^{182}W. That is why the relation between isotopes ^{182}W and ^{184}W depends on contribution of the radiogenic isotope ^{182}Hf.

The data on lunar isotopic composition have been repeatedly reconsidered. In particular, some early studies failed to take into account the influence of the nuclear reaction of cosmogenic neutrons with ^{181}Ta that results in the formation of ^{182}W (Leya et al., 2000).

The latest and most reliable data show that isotopic composition of ^{182}W/^{184}W on the Moon is almost indistinguishable from that on the Earth (Touboul et al., 2007).

The average for lunar samples $\varepsilon^{182}W = 0.09 \pm 0.1$. These authors stated that they did not find $^{182}W/^{184}W$ variations within the lunar mantle. At the same time, there are considerable variations of W isotopic composition in meteorites. We will postpone till Chapter 7 a detailed study of the Hf–W system in relation to the lunar genesis issue.

4.6.6 Magnesium

Mg isotope composition for both the Earth and the Moon samples varies within a narrow $\delta^{26}Mg$ range $\pm 1\%_o$ (Esat and Taylor, 1992; Warren et al., 2005).

The average $\delta^{26}Mg$ for the lunar upper mantle is estimated to be $-0.24\%_o$, which is almost the same as the average value for the Earth ($\delta^{26}Mg = -0.25 \pm 0.07\%_o$, Sedaghatpour et al., 2012).

The isotope fractionation is a mass-dependent. Both lunar and terrestrial samples fall on a single mass-dependent fractionation line $\delta^{25}Mg = 0.505\, \delta^{26}Mg$.

In the case of Mg the isotopic composition of the Moon and the Earth approach chondritic Mg isotope composition ($\delta^{26}Mg = -0.28 \pm 0.06\%_o$, Chactabarti and Jacobsen, 2011). However, Wiechert and Halliday (2007) argue that differentiated bodies including Earth and Moon are slightly isotopically heavier in Mg compared with chondrites. We just stress here that there is **no systematical difference in Mg isotope composition and fractionation between the Earth and the Moon.**

4.6.7 Litium

The average composition of the Moon is estimated to be $\delta^7Li = +3.8 \pm 0.4\%_o$ (Magna et al., 2006). This value is identical to δ^7Li values found for the Earth mantle olivins $+3.6\%_o$ and $+3.8\%_o$. As in the case of Mg isotopes the δ^7Li values for chondritic material are slightly lower. Magna et al. (2006) concluded that "**the Li isotope composition of Earth and Moon are identical**".

It follows from this chapter that:

Firstly, there are significant chemical differences between the Earth and the Moon including much lower iron content on the Moon, depletion of volatiles and enrichment in refractory components compared to Earth.

Secondly, isotope compositions of the Earth and the Moon are to a large extent similar, i.e., they are composed of genetically similar materials.

And, thirdly, the process that led to changes in the chemical composition of the Moon was not accompanied by isotopic fractionation.

This conclusion must be considered as the most important constraint on a model of the Moon origin.

Chapter 5

Hypotheses on the origin of the Moon

5.1 Early hypotheses

Early hypotheses regarding the origins of the Moon were mainly based upon astronomic observations. It was only when data on the composition and structure of the Moon became available that geological, geochemical and geophysical arguments appeared.

The earliest hypotheses were put forward at the beginning of the 17th century, when the heliocentric theory was just gaining recognition, and when detailed astronomic observations of the Moon were made.

It was at that time that the ideas of Descartes appeared (Descartes, 1664). He believed that satellites originated from a swarm of particles that were captured by a planet revolving around the Sun.

In the 18th and 19th centuries discussions on the genesis of the Moon focused upon ideas about secular acceleration and secular variation of the eccentricity and inclination of the Moon's orbit. A number of famous astronomers were involved in these discussions (Halley, 1695; Laplace, 1796; Adams J. C., 1859).

Immanuel Kant (Kant, 1754) appears to have been the first to suggest that tidal dissipation would retard the Earth's rotation. J. Mayer (1848) suggested that tidal interaction with Earth increased the Moon's distance from Earth. It also become clear that tidal forces acting on the Moon's rotation could have synchronized its rotation period with the orbital period, thus explaining why the same side of the Moon always faces the Earth.

Traced backward the idea of increase of the Earth-Moon distance led to the conclusion that there could have been a time in the past when the Moon and Earth were in close contact. Building upon this idea, George Darwin (1879) suggested that the Moon separated from the Earth while the latter was in molten condition.

There are certain ideas that, although not quite correct, gain a wide recognition thanks to timely support rendered by renowned scholars.

George Darwin was just as lucky as Charles Darwin, his father, had been shortly before in 1854 when he published his famous work 'The Origin of Species'. Recognition and popularization of that book were promoted by Ernst Haeckel (1834–1919), an indefatigable enthusiast of Darwinism and one of the major nature researchers and philosophers of that time.

Similarly, Geoge Darwin's idea was supported by the reputable astronomer and geologist Robert Ball (Ball, 1882). A dubious aspect of the hypothesis of G. Darwin was whether the Earth's moment of rotation would be sufficient for centrifugal separation of the lunar weight. G. Darwin was supported by the great physicist William Thomson (Lord Kelvin) who pointed out that solar tide might come to resonance with free oscillations of the molten Earth, producing distortion sufficient to disrupt the body, in spite of the deficit in angular momentum.

Geologists suggested that the Moon had been ejected from inside the Earth, emerging at a location in the Pacific Ocean that corresponded in size to the dimensions of the Moon.

The G. Darwin "fission" hypothesis was widely accepted, and remained a dominant theory of the origin of the Moon during about half of a century.

However, eventually it was recognized that Darwin's model is incorrect. American geophysicist Jeffreys (1930) has shown that the viscosity of the Earth's mantle in any realistic conditions would dampen a resonant vibration, thereby preventing the fission.

H. Urey (1952) believed that the Moon was a relic of the primary material the planets were made from, which had avoided collisions to become a satellite of the Earth. The H. Urey model was a variant of the capture hypothesis, which was also suggested and substantiated by some other scholars (Mc Donald, 1966; Singer, 1986).

There is zero probability of capturing a body approaching the Earth at high speed. Even capturing a body that emerges near the Earth and has a low relative velocity is unlikely, and would require at the time of capture on a variable orbit unclear conditions of fast energy dissipation (Wood, 1986). Besides, even if that were possible, it would be hard to explain, according to G. Wood, the significant differences between composition (for example, iron content) of the two bodies that have emerged under similar conditions.

Öpik (1972) suggested a model of disintegrative capture, in which a differentiated body passing the Earth was broken up by tidal forces, and part of its material (mostly silicate) was captured on orbit around the Earth. However, it has been shown that tidal disruption is unlikely during the short duration of a single pass (Mizuno and Boss, 1985).

The co-accretion hypothesis suggests that the Earth and the Moon accumulated from a swarm of small heliocentric bodies (planetesimals). Variants of this concept have been developed by Ruskol (1960, 1972), Harris and Kaula (1975), Wiedenschilling et al. (1986). It appears to be dynamically realistic and complies with the general theory of planet accumulation via collision of planetesimals. The co-accretion mechanism, however, does not explain high angular momentum in the Earth-Moon system (Wood, 1986). The second weak point is that both the Earth and the Moon, if they were formed by means of co-accretion, would have the same composition. Since it has been known for a long time that the Moon contains less iron than the Earth, authors developing the co-accretion concept suggested different schemes of iron splitting between the Earth and the Moon. Ruskol (1972) argues that the iron deficiency of the Moon is a consequence of the difference in strength and ductility between silicate material and metallic iron. Therefore during collision and fragmentation brittle silicates would give smaller and lighter fragments than iron and hence the proto-lunar disk would capture preferably the silicate particles. Such segregation sounds unlikely (Wood, 1986). Moreover the Moon and Earth are different not only in iron content. The Moon for instance, is enriched in refractory elements relative to Earth. This difference is even harder to reconcile with the idea of co-accretion than its deficit in iron.

Developing the idea of E. Ruskol, S. Weidenschilling (Weidenschilling, 1986) suggested that differentiation of iron and silicates in the proto-lunar disk would be more efficient if the swarm around the Earth consisted of large differentiated bodies the size of an asteroid, with silicate mantles and a metallic core. Besides, collisions of such bodies with each other or with the Earth would lead to enriching the protoplanetary disk with the substance of the Earth mantle.

There have been a lot of other hypotheses, some of them quite exotic. Recently, for example, a group of scholars suggested that the lunar substance had been thrown out of the Earth's mantle to a near-Earth orbit by means of a tremendous nuclear explosion emerging from deep inside the planet (van Westerenen et al., 2012).

However, starting from the mid-1980s, all the earlier hypotheses and even those that would be suggested later on were ousted by the mega-impact theory. It did not take long to gain almost universal recognition, and has prevailed ever since among scientific concepts of the lunar genesis.

5.2 Giant impact concept and its weaknesses

In the mid-1970s, two groups of American researchers (Hartman and Davis, 1975; Cameron and Ward, 1976) proposed a hypothesis of the impact origin of the Moon. According to this hypothesis, proto-Earth collided with another planet-sized body (with the mass of Mars) in the final stage of the planet formation process. This catastrophic event resulted in ejection of huge amounts of molten material from the Earth's mantle into a circumterrestrial orbit, where it rapidly accumulated as the Earth's satellite Moon.

Computer calculations demonstrated dynamic feasibility of the giant impact scenario (Melosh and Sonett, 1986; Benz and Cameron, 1990). The current theory of accumulation of planets (Safronov, 1969: Wetherill, 1985) admits that hundreds of bodies with masses larger than those of the Moon and Mars could have occurred during the final stage of planet accumulation in the near-solar environment, and collisions between them cannot be regarded as a unique event.

The catastrophic collision explained the high angular momentum of the Earth and the inclination of the Earth's axis to the ecliptic. The deficit in iron in the Moon could also be readily explained, because the hypothesis postulated that the collision had occurred after the formation of the Earth's core. The iron was concentrated in the core and the Moon was formed mainly from the material of the Earth's mantle.

More comprehensive study of the impact dynamics showed that the molten material ejected into orbit by a giant impact derived mostly from material of the impactor rather than from the Earth's mantle (Melosh, 2000; Canup, 2004). Development of the model allows elaboration of the impact conditions: the mass of the impactor must have been about 0.11–0.14 of the mass of the Earth, its velocity relative to Earth about 4 km/s, and the collision angle near 45° (Canup, 2004; Canup, 2008).

In its former version the mega-impact hypothesis was welcomed by geochemists (Ringwood, 1986; Wanke and Dreibus, 1986; Warren, 1992), as many geochemical features pointed to affinity between mantle materials of Moon and Earth, most importantly the three identical oxygen isotopes ($^{19}O/^{17}O/^{18}O$) of Earth and Moon. However, if the Moon derived mainly from an impactor of unknown composition the geochemical arguments in favor of the impact hypothesis are invalidated.

As to the iron problem, the new scenario of the giant impact suggested that the impactor was a differentiated body. Therefore when the impactor was destroyed during collision its metallic core drained into the proto-Earth interior (Cameron, 2000), and the Moon was produced from the impactor's mantle.

However, the silicate part of the impactor and the Earth's mantle have different cosmochemical histories. It seems unlikely that the Earth and its impactor independently evolved the same isotope composition of the major chemical elements: oxygen (Wiechert et al., 2001), silica (Georg et al., 2007), chromium (Shukolyukov and Lugmair, 2000), tungsten (Touboul et al., 2007), titanium (Zhang et al., 2011).

To explain the identical isotopes the idea of post-impact homogenization of the proto-lunar disk with the Earth's mantel has been coined (Stevenson, 2005; Pahlevan and Stevenson, 2007). It also has been suggested that the impactor could have originated from the same cosmogenic zone as the Earth – for example from one of the Earth's Lagrangian points (Belbruno and Gott, 2005).

These and the related models have been discussed in scientific literature for about 30 years (Benz et al., 1986; Stevenson, 1987; Canup and Esposito, 1996; Cameron, 2000; Canup, 2000, 2004; Reufer et al., 2011; Pahlevan and Stevenson, 2007; Nakajima and Stevenson, 2012; Salamon and Canup, 2012). However, the problems still exist.

One more snag with the giant impact hypothesis is the absence of evidence for the isotopic fractionation of elements lost as volatiles. Volatile loss has to be accompanied by a kinetic isotope effect (Humayun and Cassen, 2000). However, no isotopic shift has been detected between the lunar and terrestrial materials.

There are uncertainties and difficulties in interpretation of Hf-W, U-Pb systems as well.

We will return to these issues in Chapter 7.

Its solution of the problem of the abnormally high angular momentum of the Earth was one of the two stated achievements of the mega-impact hypothesis, the first being its explanation of the low iron content on the Moon.

That the Earth-Moon system gained excessive angular momentum as a result of the mega-impact appeared the most logical and simple explanation of its abnormal angular momentum.

It is possible, however, that the problem may have a completely different explanation. Section 1.5 contains certain data regarding specific angular momentum values of the planets. Most objects of the solar system lie on the line that is close to the rotation instability line, just like the fast rotating stars. The Earth-Moon system, too, appears to be on this line of normal angular momentum values. The solar system as a whole lies on the same line. But Mercury, Venus and Mars lie outside the line. So does the Earth, if one considers it without the Moon, and so does the Sun, too, if one considers it without the planetary system. Mercury, Venus and Mars have no satellite systems. Would it not be more logical to believe that these planets have lost the normal angular momentum they initially had? The Earth, however, kept it, as it kept the Moon. The reason for Mercury and Venus' loss of angular momentum may lie in the interaction between their parent planets with the Sun, while, in the case of Mars – in the interaction of its parent planet with Jupiter.

Chapter 6
The model of evaporative accretion

The above difficulties, some of which were detected recently while others have long been recognized, prompted considerations of possible alternative models.

An alternative hypothesis was proposed by Galimov (1990, 1995, 2004) and Galimov et al. (2005). Its essence is that the Moon was formed simultaneously with the Earth as a fragment of a double system rather than as a result of a giant impact, and that evaporation played a decisive role in the process.

6.1 Two possible paths of evolution of the solar nebula

The solar system was formed as a result of the collapse of an interstellar gas-dust cloud, which was evidently triggered by a supernova explosion. The new star (Sun) formed very rapidly, within one million years, which resulted in the preservation of short-lived isotopes produced during nucleogenesis (Podosek and Cassen, 1994). In particular, most of the ^{26}Al with a half-life of only 730 thousand years was retained.

The history of the protoplanetary cloud surrounding the Sun dates from the appearance of the earliest solid objects. These are refractory globules enriched in Ca and Al, which were first studied in the Allende meteorite and called CAI (Calcium–Aluminum Inclusions). Their age is tightly constrained (Amelin et al., 2002; Bouvier et al., 2008) as $4,567.5 \pm 0.5$ Ma and is considered to be the age of the solar system.

Chondrules were formed within 1.7–2.0 million years after the beginning of the solar system. The accretion of chondrites was completed 2–4 Ma after this. Chondrites of different petrologic classes were formed sequentially in time (Kurahashi et al., 2008; Kleine et al., 2008).

The most primitive matter of the solar system is represented by CI carbonaceous chondrites (Ivuna, Orgueil, and Murchison). Their compositions approach solar element abundances most closely. Important geochemical ratios, including Rb/Sr and Hf/W, are identical in carbonaceous chondrites and the solar nebula (Allegre et al., 2008). They are rich in carbon and contain organic matter, including amino acids and hydroxy acids, hydrous minerals (hydrosilicates), and carbonates. They were probably produced by the agglomeration of the dust component of the interstellar gas-dust nebula, the collapse of which resulted in the formation of the Sun. It did not undergo any significant temperature influence during the formation of the Sun, except for minor hydrothermal alteration. Carbonaceous chondrites of other classes (CM, CO, and CR) are mixtures of chondrules and a thermally metamorphosed matrix.

Meteorite parent bodies were 10–100 km and larger in size. The interior parts of large bodies underwent melting and differentiation. The application of Hf–W thermometry showed that maximum heating was reached by approximately 6 Ma (Kleine et al., 2008). Then the bodies cooled owing to the decay of short lived isotopes, the heat production of which (in particular ^{26}Al) sustained melting (Carlson and Lugmair, 2001). Fragments of such asteroidal materials were sources of achondrites. Within 5–10 Ma, large plyanetesimals transformed into differentiated bodies (Horan et al., 1998).

According to the standard theory, the subsequent collision of planetesimals resulted in the accumulation of increasingly larger bodies, until four planets (Mercury, Venus, Earth, and Mars) eventually formed in the inner part of the solar system. The theory of solid-phase accumulation (so called, standard theory) was developed both in Russia (Safronov, 1969) and in western countries (Harris and Kaula, 1975; Harris, 1978; Stewart and Kaula, 1980; Wetherill, 1980; Wetherill and Cox, 1985).

The hypothesis of a giant impact, i.e., the formation of the Moon owing to the collision of two planet-sized bodies at the final stage of the solid-phase accumulation, is in line with this accepted concept of planetary formation.

Thus the early formation and growth of solid bodies in the solar system are beyond doubt and can be considered as observed facts. The meteorite materials available for study were formed during the first millions of years of the existence of the solar system. The ages of respective events are accurately constrained. Meteorites themselves are products of disintegration of larger bodies of asteroidal sizes. Vestiges of this process can be observed in the numerous impact craters on the surface of planets and asteroids.

However, our suggestion, which we rely on, is that this was probably not the only path of evolution of protoplanetary material. Another way is accumulation of gas-dust clumps.

The standard theory suggests precipitation of dust particles in the central plane of the solar nebula resulted in the formation of a thin, dense sub-disk, the dusty material of which probably gave birth to solid-state evolution (Goldreich and Ward, 1973; Vityazev et al., 1990). But the remaining part of the nebula may evolve differently. The lifetime of the solar nebula is of the order of 10^7 years (Podosek and Cassen, 1994). Observations of extrasolar planetary systems indicate that stars older than several millions of years are usually devoid of visible gas-dust disks (Haish et al., 2001; Classen et al., 2006; Chambers, 2004; Camming et al., 2002).

It is possible that the solar nebula decomposed to local clumps, which could assemble to large but, in the scale of the solar system, rather compact gas-dust bodies.

The possible formation of gas-dust clumps has been discussed by many authors, starting with the pioneering study by Gurevich and Lebedinskii (1950). As early as in the 1970s, Eneev (1979) and Eneev and Kozlov (1977) produced a numerical simulation of the formation of planets and satellite systems from a gas-dust state.

E. M. Galimov (1995) suggested applying this mechanism to the formation of the Earth-Moon system. Collisions and accumulation of gas-dust clumps would eventu-

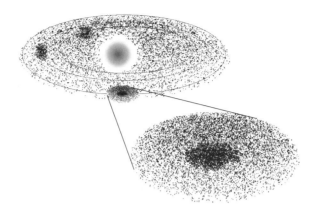

Figure 6.1. On the model of accumulation of gas-dust supraplanetary formation.

ally result in the formation of a large (within the Hill sphere) protoplanetary gas-dust formation (Fig. 6.1).

Our model postulates the formation of the Moon and Earth during the contraction and fragmentation of such a cloud. The particles are heated and subjected to evaporation during this process. We show that this model concordantly explains: the observed deficit of iron in the Moon; formation of the double Earth-Moon system from a common source and hence their identical isotopic compositions, loss of volatiles without isotope fractionation, and enrichment of the Moon in refractory elements.

We argue that evaporation played a key role in all these processes. Evaporation determines not only the chemical properties, but also the dynamics of contraction of the cloud. Evaporation from the surface of particles counteracts the gravitational attraction between particles in a cloud, and thus results in its rotational instability under smaller angular momentum than when there is no evaporation. It is well known that the angular momentum of the Earth-Moon system by itself is not sufficient for rotational instability. Due to evaporation, fragmentation of the cloud and formation of a double system occur under the real angular momentum of the Earth-Moon system.

6.2 Introduction to the dynamic model

Here we introduce our dynamic model in a simplified form (Galimov and Krivtsov, 2005) for the purposes of the subsequent chemical analysis. In the second part of the book we will present our model in more detail.

We have used the particle dynamics method (Hockney and Eastwood, 1988; Krivtsov, 2007). The method consists of representing matter as an assembly of interacting particles, to which classical dynamics equations can be applied. In these equations, inertial terms are counterweighed by particle interaction forces, which are specified as known functions of particle positions and velocities.

The motion of the particles is described by the equations of Newtonian dynamics:

The particle interaction force includes: the gravitation interaction force, the repulsion force arising at the moment of particle collision, and the term taking into account the energy dissipation at collision.

Initial conditions are defined by particle position and velocities. We assumed that the initial cloud had the parameters of the Earth-Moon systems: $K = 3.45 \times 10^{34}$ kg · m^2 sec^{-1}; the mass $M = 6.05 \times 10^{24}$ kg.

The numerical calculation showed that the angular momentum of the actual Earth-Moon system is insufficient to develop rotational instability; the collapse leads to the formation of a single body. This fact is in no way unexpected. Many attempts have been made to explain the Moon's formation by its separation from the Earth. It was however demonstrated, as we discussed in Chapter 5.1, that virtually no possible scenario of this process provides the angular momentum required for the separation. Indeed, the calculation shows that while the dimensionless parameter ω/ω_s (the meaning of this ratio is defined in Chapter 8) is below 0.42 (ω_0/ω_s value for the Earth-Moon system is 0.08) no fragmentation takes place at a rotational collapse (Fig. 6.2a). At higher values of the parameter, two different-sized bodies are formed (Fig. 6.2b,c). At the value $\omega_0/\omega_s = 0.75$, the sizes of the bodies become equal (Fig. 6.2d), and at the higher values of the parameter, multiple fragmentation occurs. Hence, we seem to have obtained yet another unsuccessful scenario of the Moon's formation as a result of the rotational instability of the initial system. However, the situation changes dramatically when one takes into account the evaporation process. The evaporation flow from a particle surface produces a repulsive impulse partly compensating the gravitation attraction.

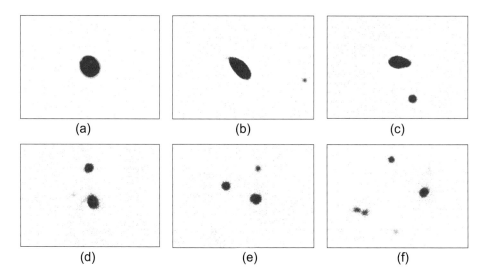

Figure 6.2. Results of computation of the rotational collapse for different values of the initial relative angular velocity ω_0/ω_s: a) 0.29; b) 0.42; c) 0.54; d) 0.76; e) 0.80; f) 0.85.

The evaporation intensity level sufficient to produce a rotational instability is of the order of 10^{-13} kg/m²· sec. The calculation, taking into account the evaporation process by including the corresponding additional term into the dynamic equation, describes a collapse accompanied by fragmentation of the cloud into two conglomerations of different sizes, which are transformed into condensed bodies. These conglomerations have a relatively elevated temperature (Fig. 6.3).

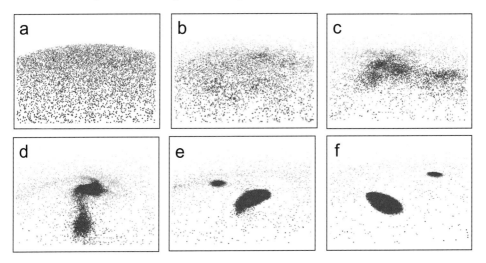

Figure 6.3. Computer simulation of rotational collapse of a cloud of evaporating particles (oblique projection). Number of particles is 10^4, $\omega_0/\omega_s = 0.70$. The spectral color scale from blue to red illustrates the temperature change from the lowest to the highest temperature in the system; a) $t = 0$, b) $t = 0.21T_s$, c) $t = 0.41T_s$, d) $t = 0.58T_s$, e) $t = 0.80T_s$, f) $t = 1.07T_s$, where $T_s = 2\pi/\omega_s$.

6.3 Loss of iron and enrichment in refractories

The most interesting aspect of our hypothesis is that the heating of particles during the collapse provides a possible explanation for the iron loss from the Moon.

We do not proceed in our model from any preconceived chemical composition. However, as we claim evaporation to be a key factor of modification of primary composition of the pre-lunar material, the elevated concentration of refractory elements is in logical agreement with our model.

As the Moon is enriched in the rock-forming refractory elements Al, Ca, and Ti, it must be depleted not only in minor elements like Na, K, Rb, Cs, In, Bi, Tl, Pb etc., but in Si and Mg as well, which are usually not regarded as volatiles. This feature is not evident from comparison of the chemical composition of the Moon, Earth and CI chondrites as shown in Table 4.3 (Chapter 4).

The relative depletion or enrichment in a particular component is established by normalizing its content to some component whose abundance is assumed to be constant. Silicon is often used as a normalizing element. But in such a case, the fact of Si depletion in the Moon would be disguised. As refractory elements do not fractionate during evaporation one of them may be used as a normalizing component.

Table 6.1 shows the Al-normalized abundances of elements in the compositions of the Moon, the Earth and Cl chondrites.

Table 6.1. Al-normalized elemental compositions of CI, the Earth's mantle, and the Moon.

Element	CI	Earth	Moon: *
Si	12.3	12.0	6.4
Ti	0.05	0.05	0.05
Al	1.0	1.0	1.0
Fe	21.5	24.7	3.2
Mg	11.0	10.9	6.1
Ca	1.04	1.05	1.02
Na	0.57	0.13	0.02
K	0.06	0.01	0.0025

* Following S. Taylor's (1986) Moon composition.

It is seen from Table 6.1 that the Moon is depleted in not only typical volatiles but also in such moderately volatile elements as Si and Mg.

The normalized concentrations of the refractory elements Ti and Ca in all three objects are practically identical.

Since the publication of the well known condensation sequence (Grossman and Larimer, 1974; Larimer, 1979, 1986), iron has been considered as a refractory element. Indeed, the first condensed substances are Ca, Al, and Ti oxides, then Fe and Ni. Silicate and oxides of Mg and Si are formed later, followed by the typical volatiles K and Na, and the even more volatile Rb and Pb (Fig. 6.4).

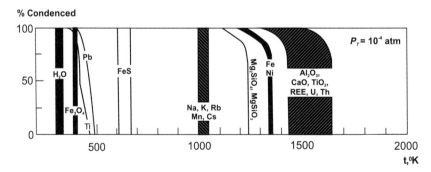

Figure 6.4. Sequence of element condensation from a high-temperature gas according to Larimer (1986).

Thus Fe and Ni are estimated to be early condensates immediately following the most refractory components, while the main minerals of Mg and Si (forsterite and enstatite) are condensed after these. This is probably the reason why iron has never been considered as a volatile component (except in a footnote in Wood's (1986) paper).

However, if the evaporation of carbonaceous chondritic material, enriched in iron oxide, is considered instead of high temperature gas-to-solid condensation (CAI and chondrules), it appears that iron can be lost as FeO. Ferrous iron is not volatile by itself. It is decomposed to elemental iron and oxygen, and iron vaporizes from a silicate or oxide mineral more efficiently than from a pure metallic phase (Hashimoto, 1983).

Figure 6.5 shows the calculated curves of the equilibrium partitioning of each element of the system SiO_2-TiO_2-Al_2O_3-FeO-MgO-CaO-Na_2O-K_2O-H_2O-C between solid and gas phases at temperatures varying from 500 to 2,300 K (Galimov, 2004). The initial proportions of the components corresponded to the CI composition (carbonaceous chondrite Orgueil) and the pressure was $P = 10^{-3}$ bar. All the elements were initially incorporated in the solid phase. It is seen that Fe begins to partition into the gas phase at a temperature lower than that under which Si and Mg begin to evaporate. This result is in agreement with the experimental data.

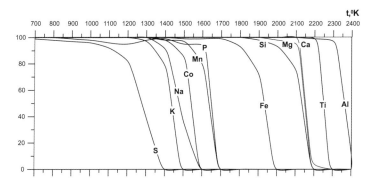

Figure 6.5. Theoretical curves of element partitioning between solid and gas phases for the multicomponent system corresponding to the composition of carbonaceous CI chondrites, SiO_2-MgO-Al_2O_3-TiO_2-CaO-FeO-K_2O-Na_2O-H_2O-C with addition of Mn, P, and Cr compounds, at various temperatures and a pressure of $P = 10^{-3}$ bar. The diagrams were computed by Yu. I. Sidorov and V. E. Kulikovskii using the Infotherm program and thermodynamic database compiled at the Vernadsky Institute of Geochemistry and Analytical Chemistry, Russian Academy of Sciences (from Galimov, 2004).

Figure 6.6 summarizes the experimental results of various authors, who studied variations in element concentrations during melt evaporation in vacuum (de Maria et al., 1971; Yakovlev et al., 1972; Hashimoto, 1983; Markova et al., 1986). Ka_2O and Na_2O are the first rock-forming components to escape from the melt. Then the melt is depleted in the sequence iron, silicon and magnesium. In the course of evaporation, the melt becomes enriched in Al, Ca and Ti. Ferrous iron is next to K_2O and Na_2O with respect to volatility.

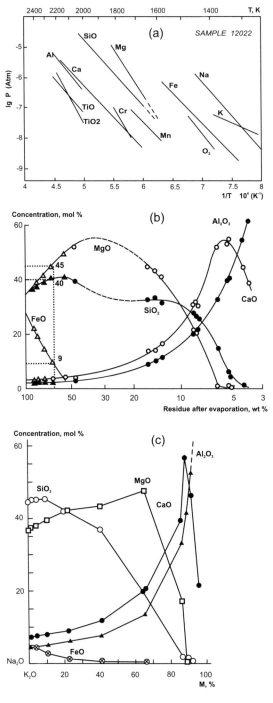

Figure 6.6. Experimental data for volatile loss during melt evaporation in vacuum according to the data of (a) de Maria *et al.* (1971), (b) Hashimoto (1983), and (c) Markova *el al.* (1986).

The experimental data presented in Table 6.2 reveals close quantitative similarity between the composition of chondritic material after evaporation of about 40 % of the initial mass and the Moon composition. For convenience of comparison the data presented in Fig. 6.5b in mol % are recalculated in Table 6.2 into weight %. At least for the studied element one can see the outstanding similarity between the composition of the Moon and Cl-material subjected to 40 % vaporization.

Table 6.2. Comparison of composition of the Moon with that of the residue after evaporation of 40 % of primary material of chondritic composition (weight %).

Component	Primary composition (CI-chondrite)	Composition[*] of the residue after 40 % evaporation	The Moon
MgO	23.5	31.9	32.0
SiO2	35.0	42.9	43.7
FeO	36.9	15.8	13 + Fe in the core
Al2O3 + CaO	4.6	9.4	10.8

[*] The experimental data from Fig. 6.5b

Thus the evaporation mechanism considered above provides a basis for a fundamentally different interpretation of the iron deficit. The iron content in the Moon decreases due to evaporation concordantly with other evaporating elements down to the values observed (estimated) for the Moon.

6.4 Asymmetry of accumulation of Earth and Moon

Fragmentation results in the formation of two high temperature fragments: the embryos of Earth and Moon. The question arises as to why the content of iron should be so dramatically different in the two fragments, one of which is destined to become the Earth and the other the Moon, since one would expect that both the bodies formed by fragmentation of the cloud should be equally depleted in iron. The solution of the problem is asymmetric growth of the embryos (Galimov et al., 2005; Le-Zaharov and Krivtsov, 2009).

An important feature of collapse with a chaotic component of particle velocity is that a significant fraction of particles stays scattered over space after the condensed bodies are formed, and the temperature of these scattered particles is substantially lower than that of the condensed bodies. Later the scattered particles are accumulated by the condensed bodies, which takes several orders of magnitude more in time that the process of formation of the condensed bodies (Fig. 6.7).

As mentioned above, after the formation of condensed bodies, the environment still contains a large quantity of dust materials that are deposited onto the fragments which have already formed.

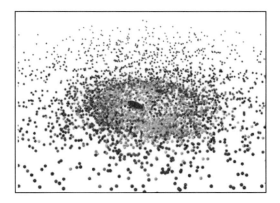

Figure 6.7. The cloud of particles in space surrounding the condensed bodies immediately after their formation ($t = 1.07T_s$).

The computer simulation demonstrates that if the masses of the bodies are different, the larger body grows faster. This dependence can be closely approximated with a square function. Therefore, a random initial difference of masses leads to a situation where the smaller body does not significantly gain in mass, while the larger one accumulates most of the initial particle pool (Fig. 6.8).

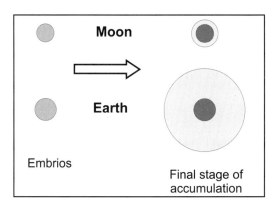

Figure 6.8. Asymmetry of accumulation of Earth and Moon.

For example, in the case of a 4-fold difference between the initial mass of the two fragments, the larger of them (future Earth) increases its mass 26 times, while the smaller one (future Moon) only by 30 %. The dynamics of this process is considered in detail in subsequent sections (Part II, Chapter 9).

Within the framework of the study of collapse dynamics the temperature conditions in the compressing gas-dust cloud have been calculated. The initial temperature was assumed to be 300 K, and the average heat capacity of the solids 800 J/kg K.

Section 6.4 Asymmetry of accumulation of Earth and Moon

The results are shown in Figs. 6.9–6.10. In the initial period of collapse temperatures of both the Earth's and the Moon's embryos slightly exceed 2,000 K. This temperature is sufficient for vaporization of volatiles, including FeO, but not refractories. In the next period of time the temperature increases, approaching $\sim 4,000$ K in the interiors of the embryos. However, at the surfaces of the consolidated bodies the temperature remains under 2,000 K.

And in the surrounding dust material it remains under 750 K.

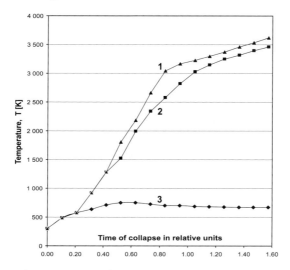

Figure 6.9. Change of temperature during collapse. 1 – Earth's embryo, 2 – Moon's embryo, 3 – surrounding cloud of particles.

Thus, initially the high-temperature embryos of the Moon and the Earth were equally depleted in iron. Later on, both the Moon and the Earth acquired the colder materials of the residual part of the cloud. However, the Moon accumulated little, maintaining the iron deficit, whereas the Earth embryo collected the major fraction of the surrounding material, whereby the composition of the Earth came closer to the composition of the cloud as a whole.

We have assumed that the parental gas-dust cloud of the Earth and Moon was chemically similar to the solar nebula. The dust component approached the composition of CI carbonaceous chondrites (in particular, iron occurred as FeO), whereas the gas medium was dominated by hydrogen, similar to the solar nebula.

Hydrogen executes two functions. First, it is a carrier gas providing the hydrodynamic removal of compounds and elements migrating into the gas phase during the compression of the gas-dust cloud. Second, it is an efficient reducer (Fig. 6.11).

Temperatures sufficient for FeO evaporation will be reached in the central part of the fragmenting cloud. This is our starting assumption. The evaporation rate should be small on the time scale of fragmentation (10^5–10^6 years).

58 Chapter 6 The model of evaporative accretion

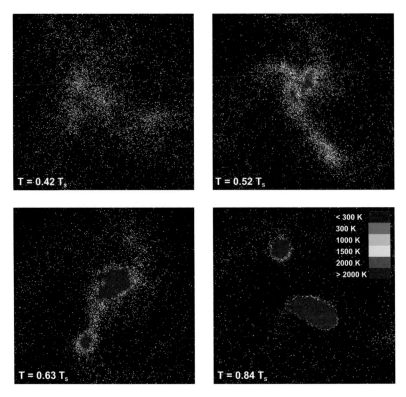

Figure 6.10. Temperature distribution in consecutive steps of compression of the collapse.

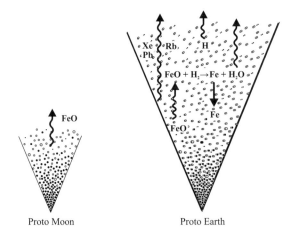

Figure 6.11. Segregation of metal and the hydrodynamic escape of volatiles in the model of the formation of the Earth and Moon from a supraplanetary gas-dust body.

Experiments showed that the rate of FeO evaporation is $1.14 \cdot 10^{-5}\,\text{g}\,\text{cm}^{-2}\text{s}^{-1}$ at 1,900 K (Kazenas and Tsvetkov, 2008). At such a temperature, half of the FeO will evaporate from a particle of 1 mm radius within $0.2\cdot 10^4$ s. Obviously, temperatures did not exceed 1,000 K. There are no experimental data for temperatures below 1,000 K.

The evaporation of FeO is incongruent. In fact, Fe and O_2 occur in the vapor. The oxygen is fixed by hydrogen, and the metallic iron is separated in a dispersed state.

Reducing FeO produces H_2O. As H_2 dissipates H_2O becomes the main component of the gas phase and main carrier gas. The process of evaporative accretion passes from gas-dust regime into particle-vapor regime.

The embryos of the Moon and the Earth are depleted in Fe owing to FeO loss. However, during the continuing accretion of the Earth occurring in the hydrogen medium, the dispersed iron will be condensed into metallic particles. The solid metallic particles can precipitate against the gas flow toward the center of mass. Thus, the formation of the Earth's future core begins as a result of the precipitation of metallic iron through the vapor-gas phase. The Moon retains residual iron mainly as FeO, whereas the Earth gains metallic iron during further accretion.

It has always been a difficult problem to explain why the Earth and other planets contain massive metallic cores, whereas oxidized iron occurs in the primary matter. The primary material of carbonaceous chondritic composition contains iron only in the FeO form. Ordinary chondrites contain very minor amounts of metallic iron. The formation of a massive metallic iron core requires the removal of the equivalent mass of oxygen. No adequate mechanism has been proposed for this process. Iron reduction by primary hydrogen during the compression of the particle cloud provides a natural solution to this problem. The water produced by this process is squeezed together with hydrogen $+$ H_2O in a vapor-gas phase from the compressed cloud of particles and expelled from the proto-Earth cloud. This is the process of hydrodynamic escape.

It is important to note in this connection that the general problem of Earth degassing is interpreted in a new manner. The Earth not only lost the volatile compounds of light elements, such as water, nitrogen, methane, and CO_2, but also became depleted in heavy volatiles, which should have been gravitationally retained. The hydrodynamic conditions in a hydrogen (water) flow result in the removal of heavy volatile elements, including Xe and other noble gases, as well as non-atmophile heavy elements, such as Rb and Pb, occurring in a vapor phase under the P-T conditions of the vapor-dust medium.

Chapter 7

Geochemical constraints and how the giant impact and evaporative accretion concepts satisfy them

7.1 Identity of isotope compositions of the Earth and the Moon

Similarity of the isotope composition of the Earth and the Moon is the strongest constraint on the hypotheses of the origin of the Moon.

Such a similarity is a natural consequence of the formation of the Earth and Moon from a common source, as is suggested by the evaporative accretion concept.

On the other hand this is the main difficulty of the giant impact concept. There exists significant heterogeneity in the solar system with respect to isotope distribution. The coincidence of $^{16}O/^{17}O/^{18}O$ isotope fractionation lines for the Earth and Moon is often discussed. But as shown in Chapter 4 many other elements reveal surprisingly identical isotopic composition for Moon and Earth despite the significant range of cosmochemical variations between the two. This is in fact quite astonishing, considering that finding even occasional compositional coincidence between two genetically unrelated cosmic bodies is as unlikely as meeting two people with identical finger-prints.

Various suggestions have been made to save the hypothesis; for instance, it was supposed that the proto-lunar material occurred in the atmosphere of silicate vapor ejected from the Earth by the giant impact, and the oxygen isotopic compositions of the Moon and the Earth were equilibrated (Stevenson, 2005).

Even if one admits homogenization of oxygen isotope composition in the atmosphere of silicate vapor (which is, in fact, difficult to imagine), it is impossible to apply this mechanism to isotope homogenization of such refractory elements as Ti and W (see Chapter 4).

But the most important objection against the homogenization idea is the fact that the Moon and the Earth in spite of identical isotope composition are significantly different in their chemical compositions.

Isotope homogenization will inevitably "result in a silicate Moon [that is] isochemical with the terrestrial mantle" (Pahlevan and Stivenson, 2007). However, the Moon is depleted in volatiles and enriched in refractory elements. There is a rule in isotope chemistry: isotope equilibrium is achieved later and as a result of chemical equilibrium. In other words, isotope homogenization in chemically heterogeneous material would inevitably erase the chemical differences.

Another suggestion was that the Earth and the impactor were formed in the same zone of the solar system, practically on the same orbit, and inherited identical oxygen isotopic ratios from this zone. One variant of this hypothesis is the accumulation of the material of the impactor in the Lagrange points (Belbruno and Gott, 2005). The stable accumulation and occurrence in a Lagrange point of a massive planetary body seems unlikely. Pahlevan and Stevenson (2007) have noted that according to the current theory of accumulation of planets (Chambers, 2004), even if the impactor came near Earth's orbit at one of the Lagrange points a compositional difference between the Earth and the impactor may be expected.

Attempts were made to modify the conditions of mega-impact so as to enlarge the relative contribution of terrestrial material in the proto-lunar disk (Reufer et al., 2011). The following models have been tested: 1) an iron-enriched impactor similar to the planet Mercury (30 % rock, 70 % iron); 2) an ice-enriched impactor (50 % water ice, 38 % rock, 12 % iron) simulating an object born beyond the snow line of the early solar system in the giant gas-planets zone; 3) a more head-on collision with impact velocity above the escape velocity, putting a higher fraction of the terrestrial mantle material into orbit. Computer modeling has shown that not one of those models provides a satisfactory solution. The most promising case is a 30° impact angle collision with an impactor about 0.2 of the Earth's mass and 1.3 times the escape velocity, leading to a disk of roughly lunar mass with the mantle derived fraction 0.6 (Reufer et al., 2011).

Thus the coincidence of isotope characteristics of the Earth and the Moon, especially $^{16}O/^{17}O/^{18}O$ composition as well as $^{46}Ti/^{47}Ti/^{50}Ti$, $^{53}Cr/^{52}Cr$ $^{182}W/^{184}W$ and other ratios, pose difficult, in our opinion, insurmountable problems for the giant impact hypothesis.

7.2 Loss of volatiles without isotope fractionation

The problem of volatiles in connection with the Moon includes two questions. (1) How could the Moon retain some high volatile components, including water, while being significantly depleted in volatiles, including even Si, Mg and Fe? (2) Why is the loss of volatiles not accompanied by isotope fractionation?

These questions are addressed to both the giant impact and the evaporative accretion concepts.

However, there is evidence that the Moon contains some volatiles in its interior. It is likely that volatiles, including H_2O and CO_2 have played a role in lunar igneous processes. The pyroclastic glasses are believed to be the product of fire-fountain eruption driven by volatile degassing (McCubbin et al., 2011; Rutherford and Papale, 2009; Sato, 1979). The discovery of water in lunar rocks (Saal et al., 2008) exacerbates the problem. We will talk about lunar water separately in the next paragpaph. There we will show that the presence of water on the Moon is of the same nature as loss of volatiles without isotope fractionation. In both cases, the reason is that the formation mechanism of the Moon has enabled the inter-phase fractionation processes to occur within a **closed** system.

This section will answer the second question.

One should bear in mind that considerable isotope fractionation has been observed during melt vacuum evaporation (Wang et al., 1994, 1999, 2001; Humayun and Clayton, 1995; Humayun and Cassen, 2000).

Isotope fractionation caused by the kinetic isotope effect takes place under the following two conditions: 1) if vapors are removed continuously from the evaporating surface, for example, in the case of vaporization into a vacuum, and 2) if the upper layer of an evaporating liquid is refreshed faster than the liquid proper evaporates, i.e., the liquid is quite actively agitated (Fig. 7.1a).

Evaporation processes are always accompanied by the kinetic isotope effect. However, final isotope fractionation is not obligatory if the rate of internal mixing is lower than the rate of evaporation. Yong (2000) has quantified this process in accordance with the ratio of evaporation rate to subsurface diffusion.

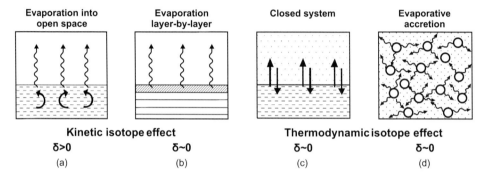

Figure 7.1. Types of isotope effects and isotope fractionation.

Furthermore, isotope fractionation might not occur when the fractionated layer is removed completely from the surface of the evaporating phase (Fig. 7.1b). One particular mechanism in this process is layer-by-layer evaporation. It is usually observed during the evaporation of solid particles in a vacuum, when the diffusion rate in the solid is lower than the rate of vapor escape. This effect was experimentally studied in the example of forsterite crystal evaporation (Davis et al., 1990; Wang et al., 1999). Though layer-by-layer evaporation suppresses isotope fractionation, lack of the isotopic changes in the Moon cannot be explained by this way as layer-by-layer evaporation suppresses chemical fractionation as well, while the Moon shows different loss of elements depending on their volatility.

In contrast to evaporation, equilibrium condensation may occur without isotope fractionation.

The reason for this is that condensation normally starts in a saturated vapor condition. In this condition equilibrium is achieved between vapor and liquid phases. This condition features a thermodynamic isotope effect, which is either small or almost absent for most elements – except for the lightest ones forming mobile phases (H, C, N, O, S) – and at higher temperatures for virtually all elements (Fig. 7.1c).

It is thought therefore that the presence or absence of isotopic effects can be used as an indicator of the type of process, evaporation or condensation, prevailing in the gas-condensed phase system (Wang et al., 1999). The fact that solar system bodies show varying degrees of volatile depletion but no isotopic effects is believed to be due to the loss of volatiles during the stage of successive condensation of the protoplanetary nebula.

Condensation processes taking place in the solar nebula have actually left no noticeable traces of isotope fractionation. Only Ca-Al-inclusions show any isotope shifts that appear to have been caused by unequilibrium evaporation.

However, the condensation mechanism cannot be applied to fractionation processes that occurred during the formation of the Moon.

It should be noted that the frequent interpretation (e.g. Jones and Palme, 2000) of isotope fractionation during condensation as being small due to the thermodynamic (equilibrium) isotope effect, and of isotope fractionation during evaporation as being significant due to the kinetic isotope effect is not correct as a general case. Actually, if **condensation** occurs from just a slightly oversaturated vapor the fractionation is controlled by the small thermodynamic isotope effect. But similarly, when **evaporation** occurs into the slightly undersaturated surrounding vapor it is also controlled by the thermodynamic isotope effect. So on the contrary: the isotope fractionation during condensation can be significant (due to the kinetic isotope effect), if condensation occurs from a highly oversaturated vapor, for example as a result of a sharp decrease in temperature. Formation of autholiths in kimberlite enriched in ^{13}C isotope as high as $+30$ to $+50‰$ in δ^{13}C terms may serve as an example of such a condensation process (Galimov, 1991). Similarly, if evaporation proceeds into a significantly undersaturated vapor phase (for example into a vacuum) the isotope fractionation is also determined by the kinetic isotope effect.

Isotope fractionation may be small or absent during evaporation if evaporation occurs in a closed system. The evaporating particles are surrounded by their vapor. Isotopic compositions of vapor and liquid are equilibrated due to isotope exchange and controlled by negligible equilibrium (thermodynamic) isotopic effects (Fig. 7.1d).

This situation is suggested by our model of evaporative accretion. One more mechanism of volatile loss without isotope fractionation is so-called hydrodynamic escape, when a vapor is swept away by a carrier gas. This mechanism is mistakenly applied to explain the absence of isotope fractionation when the issue of volatiles is considered in relation to the mega-impact hypothesis. Hydrodynamic escape prevents isotope fractionation in a gas phase as the isotope species of any vapor component entrained by the carrier-gas cannot escape of its own accord. However, if isotope fractionation occured at a vapor-liquid phase boundary, then hydrodynamic outflow would carry away a vapor component already isotopically fractionated. If the kinetic isotope effect occurred in the liquid-vapor phases boundary in the giant impact process then hydrodynamic escape would be unable to prevent eventual isotope fractionation.

Some authors have argued that during the giant impact, vapor and liquid phases could have coexisted long enough to reach isotope equilibrium (e.g. Canup, 2004). It is unknown if equilibrium isotope fractionation is achievable in the conditions and time-scale of the putative proto-lunar disk (Salmon and Canup, 2012).

The model illustrated by Fig. 7.2 meets the conditions under which isotope fractionation is not manifested: (1) hot particles are in equilibrium with their vapor; (2) the vapor is carried away by hydrodynamic flow, and (3) the vapor is expelled layer-by-layer into the outer zone of the cloud during its contraction. The squeezed vapor is swept by a permanent flow, for instance by the solar wind. This mechanism explains why the loss of volatiles occurs without any significant isotope fractionation.

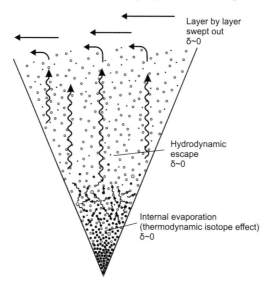

Figure 7.2. Volatile loss without isotope fractionation in evaporative accretion model.

The conclusion regarding the absence of fractionation during loss of volatiles is of central importance. It conforms with the basics of the suggested model. However, isotope fractionation in the above processes, small as it is, does not have to be zero. Therefore, further studies may yet review expected isotope effects, even if their volumes are small, provided they are within the limits of possible thermodynamic isotope effects. It is also possible that certain processes exist that deviate from the mainstream and may express themselves as a more manifest isotope fractionation.

This may include the isotope shifts observed in the isotopic composition of iron.

It was suggested in Chapter 6 that the evaporation of FeO is incongruent. The oxygen is fixed by hydrogen, and the metallic iron is separated in a dispersed state.

The iron aggregates to the particles descending to the center of mass. Thus the iron forms a separate phase, which is not in isotope exchange with the ambient medium (Fig. 7.3).

A kinetic isotope effect must operate in such a case. The metallic particles migrating to the core must be enriched in the light isotope (^{54}Fe), and the iron retained in silicate particles must be enriched in ^{57}Fe. This is consistent with observations (Poitrasson et al., 2004).

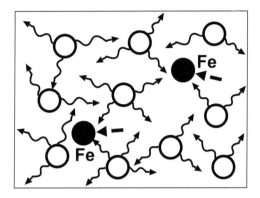

Figure 7.3. Isotope effect during irreversible formation of iron particles.

Both the Earth's mantle and the silicate phase of the Moon are enriched in ^{57}Fe (by 0.1 and 0.2‰, respectively) compared with chondrites. (Poitrasson et al., 2004; Williams et al., 2005; Weyer et al., 2005).

Poitrasson (2007) suggested that the enrichment of the Moon in ^{57}Fe could be attributed to the giant impact: "best explained by loss of light iron isotopes during the high temperature event accompanying the interplanetary impact that led to the formation of the Moon". However, this author avoided the question of why the Moon is not enriched in the heavy isotopes of other elements that were lost from the Moon as volatiles during the giant impact. However, the observed variations are in agreement with the proposed scenario for the formation of the Earth's core.

7.3 Water in the Moon. D/H ratio

Owing to the nearly orthogonal lunar rotation axis in relation to the ecliptic plane (1.54° at its current state), even shallow craters in the polar regions of the Moon are always shadowed. Due to continuously low temperatures in those areas, any volatile components, including water, are frozen out there. Ideas regarding possible concentration of water ice in the polar craters have been put forward ever since studies of the Moon began (Watson et al., 1961; Arnold, 1979). In 1994 during the Clementine radar mapping mission (Nozette et al., 1996) and in 1998 during the Lunar Prospector neutron flux recording mission (Feldman et al., 1998) the effect of enhancement of hydrogen abundance in permanently shadowed polar regions of the Moon was actually recorded.

As a result of more accurate subsequent mapping of the lunar surface during the projects Lunar Reconnaiassious Orbiter (LRO) (Mitrofanov et al., 2010, Miller et al., 2012) and Chandrayaan (Pieters et al., 2009) data regarding hydrogen anomalies on the Moon were corrected and expanded. On the LRO spacecraft the Lunar Exploration Neutron Detector (LEND) with Collimated Sensors of EpiThermal Newtrons (CSETN) provided high resolution recordings (Mitrofanov et al., 2010). Three craters, Shoemaker and Cabeus in the south polar region and Rozhdenstvensky in the north, showed significant neutron suppression from 5.5% to 14.9%. Measurements of water in the plume caused by the impact of the LCROSS spacecraft on the lunar surface at the Cabeus crater area gave readings of $5.6 \pm 2.9\%$ water content in the regolith (Colaprete et al., 2010). However, many permanently shadowed craters did not display neutron suppression (Sanin et al., 2012). Some authors question the accuracy of the data obtained by the LEND CSETN detector (Eke et al., 2012). Evidently, more accurate mapping of hydrogen presence on the lunar surface is desirable.

The NASA Moon Mineralogy Mapper (M^3) on the Indian spacecraft Chandrayan-1 measured OH and H_2O absorption wavelengths. The presence of the hydrated phase in the upper few millimeters of the lunar regolith has been established (Pieters et al., 2009). The authors believe that the surficial processes involving the solar wind are the most likely explanation of the M^3 observations.

In general three sources of water may occur: retention of water from meteorite and comet impacts, implantation of solar wind hydrogen, and degassing of indigenous water.

In contrast to absorption spectra which characterize 2–3 mm of the upper layer, neutron detection represents about 50 cm of the regolith layer. The mappings produced by these different techniques do not completely coincide (Clark, 2009; Pieters et al., 2009). Any water coming onto the lunar surface with falling comets and meteorites would have been frozen out in the cold traps of the continuously shadowed areas. The fact that the hydrogen anomaly exists both around the shadowed craters in the polar regions and on lower latitudes, in well-lighted areas, means that the registered water may be part of the Moon's own hydrogen.

Water was found in the composition of volcanic glasses (Saal et al., 2008). Its presence was so obvious ($\sim 1,5$ wt%, Greenwood et al., 2010) that questions were raised immediately regarding the high-temperature genesis of the Moon, since it seemed incompatible with the Moon's volatile super-depletion.

Lunar water appeared to be quite rich in deuterium. This means that solar origin of lunar water as a common occurrence can be excluded, since the isotopic composition of lunar water is very different from that of solar wind. Solar wind is characterized by the values $\sim \delta D \simeq -850\%$, whereas studies of water contained in lunar minerals have shown that it has mostly positive values of δD (from $+200$ to $+1,200\permil$) (Fig. 7.4).

The isotopic composition of hydrogen contained in the water of lunar glasses varies within $+179 \permil$ to $+5,420 \permil$ (Weber et al., 2011, Saal et al., 2012). These δD values are outside the δD range that describes the Earth's water (Robert et al., 2000). Water in the oceans has $\delta D = 0$ ‰, while the mantle water is, as a rule, isotopically lighter. Mantle phlogopites on the Earth have $\delta D \simeq -80$‰.

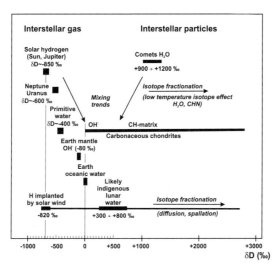

Figure 7.4. D/H ratio in lunar water and related objects.

The maximum amount of water (960–6,050 ppm) was found in apatite from lunar mare basalt (Boyce et al., 2010; Greenwood et al., 2011). The δD values of that water varied from $+391$ to $+1,010$‰.

The D/H of the mare basalts from different landing sites have been found to be similar (Boyce et al., 2010; Greenwood et al., 2011): the mean and standard deviation of δD analyses of mare basalts 10,044, 12,039 and 75,055 are $+681 \pm 132$‰ ($n = 27$); the mean and standard deviation of δD of 12,039.43 are $+698 \pm 61$‰ ($n = 9$); and the mean δD of 10,044 is $+606 \pm 30$‰ ($n = 4$). The δD of two analyses of an intrusive highlands alkali anorthosite clast are a little different from mare basalts, but also elevated relative to Earth ($\delta D = +238 \pm 72\%$ for sample 14,305 and $+344 \pm 53\%$ for sample 14,303).

Chen and Zhang (2012) measured water concentration in olivine-hosted melt inclusions in lunar basalts. Despite texture and chemical differences the studied lunar basalts contain approximately similar pre-eruptive water concentration (≥ 100 ppm).

The slightly higher content of deuterium may have been caused by the kinetic isotope effect during diffusion loss of H, and also by the process of spallation due to sun rays acting on the lunar surface. The spallation production rate of D at the lunar surface is $6 \cdot 10^{-10}$ g/10^8 years according to Liu et al. (2012). It means that during 100 Ma of exposition of soil containing about 100 ppm of water its isotopic composition would change from the solar D/H ratio ($\sim 2 \cdot 10^{-5}$) to $D/H = 0.8 \cdot 10^{-4}$. The latter is still less than the D/H ratio of terrestrial water ($\sim 1.5 \cdot 10^{-4}$). If the initial isotopic composition in the lunar soil were at the terrestrial level ($1.5 \cdot 10^{-4}$), then under the same conditions D/H in the soil would increase to $2 \cdot 10^{-4}$ or $\delta D = +330\text{\textperthousand}$. Thus significant effect of spallation is possible only if the water content is relatively small. The observed δD values of $+300$ to $+800\text{\textperthousand}$ when water content exceeds 200–500 ppm are most likely characteristic of indigenous lunar water.

The rich content of D isotope in the water may also be attributed to the kinetic isotope effect during diffusion loss of water.

The H_2/H_2O ratio is a function of oxygen fugacity. Under conditions of lunar basalts (IW-bufer) H_2 is the dominant component. The H_2 diffusion rate exceeds the H_2O diffusion rate by a factor of 10 at 1,400 °C and of about 10^4 at 400 °C (Zhang, 2011). At ~ 100 °C (the average daytime temperature on the Moon) the H_2 diffusion constant is $6 \cdot 10^{-14}$ m^2/s^{-1} with a diffusion distance of about 1.4 mm in one year, meaning that any molecular hydrogen in lunar glasses would be lost (Zhang, 2011). That means that there is a hydrogen flux in lunar rocks.

The influence of the kinetic isotope effect and the spallation effect must increase when the content of primary water is low. However, in most samples there is no connection between isotopic composition and water content.

Water in the lunar rocks may partially come from hydrogen contained in the solar wind. Implanted solar hydrogen was found in lunar samples by some earlier scholars (Epstein, Taylor, 1970). It is enriched with the light isotope ($\delta D \simeq -850\text{\textperthousand}$). Liu et al. (2012) studied the D/H in lunar agglutinates. Lunar agglutinates are small aggregates of regolith material cemented by melts generated by micrometeorite impacts. They contain nano-particles of zero valency iron that resulted from reduction of FeO by solar protons and, possibly, from impact reduction by micro-meteorites (Taylor et al., 2001; Friedman et al., 1972). The studied agglutinates had δD values from -845 to $-562\text{\textperthousand}$ with water content from 240 to 72 ppm (Liu et al., 2012), although one sample (10,084 soil from Apollo-11) displayed extremely high D enrichment ($+4,200$ to $+5,400\text{\textperthousand}$).

There is no doubt that secondary effects take place. And the above examples of anomalously high content of deuterium are probably linked with those processes.

On the whole, however, there is normally no relation between water content and isotope properties. Therefore, δD values in the range of $+300$ to $+600\text{\textperthousand}$ should be considred as typical of indigenous lunar water.

The available data prove that the Moon has preserved part of its indigenous water. Its isotopic composition lies within the same δD range as the water of comets and CI chondrites, i.e., the δD of the primitive dust substance that, according to our understanding, was the origin of the lunar substance.

The different isotopic compositions of the Earth and the Moon do not fit well with the mega-impact hypothesis.

At the same time, any difference between material or isotopic compositions of the Earth and the Moon that originated, according to the mega-impact hypothesis, from an alien impactor, should not be unexpected. However, as already mentioned, there is proof of genetic kinship between the substances of the Earth and the Moon. The only way to explain the surprising similarity between oxygen isotope fractionation lines of the Earth and the Moon, as well as the identical isotopic compositions of W, Ti and other elements, is the concept of full intermixing of the proto-lunar disk substance with that of the Earth, (Pahlevan and Stevenson, 2007). Yet, as we have noted already, this explanation is utterly improbable, since if we were to follow this logic, water would have to be the first to mix.

Some papers suggest that the δD of lunar water is little or no different from that of the Earth, and the observed enrichment of lunar water with deuterium is connected with secondary isotope effects: the kinetic isotope effect during diffusion losses and the spallation effect (Weber et al., 2011). The data available show that this is unlikely. Even if one considers possible fractionation and radiogenic addition of D, the isotopic composition of lunar hydrogen is different from that of the Earth. The samples that have the highest water content and, therefore, are least modified by any secondary processes, are those containing hydrogen with an average δD of $+340\%_o$ (Saal et al., 2012).

Moreover, the very presence of such a volatile component as water in the lunar substance proves (as with the other) that volatiles were lost as a result of inter-phase equilibrium rather than due to evaporation in an open system.

Why do isotopic compositions of hydrogen on the Moon and the Earth differ?

The simple answer to this question lies in the formation mechanism of the Moon and the Earth from the common gas and dust cloud (Fig. 7.4). During the initial formation stage of embryos of the Moon and the Earth, their water came from the dust component, i.e., from water contained in the CI chondrite and comet parts. During further formation of the Earth, there was ever-increasing contribution of water that originated from the reduction of FeO by means of the primary hydrogen in the gas and dust cloud and from formation of the metallic core of the Earth. This is the hydrogen depleted of deuterium. Isotopically light water accumulated in the Earth, which eventually led to the observed difference between isotopic compositions of the Earth and the Moon.

Since it is possible to conclude that lunar water is of autochthonous origin, i.e., it was inherited from the formation period of the Moon, the question that first emerged at the beginning of the previous section reappears: How did the Moon manage to preserve its water considering its overall depletion of volatiles, including partial loss of none too volatile components, such as Mg and Si?

The answer to this question has the same basis as the answer to another one: Why is the Moon depleted of volatiles without any manifest isotope fractionation?

It follows from theoretical calculations (for example, Fig. 6.4 and 6.5 in Chapter 6) that the volatility of a compound is confined by a narrow temperature range. In other words, by the time a less volatile component just begins to vaporize, the more volatile components have almost completely disappeared. A complete loss of volatiles occurs if the vapor is continuously removed from the surface of the condensed phase. However, if evaporation occurs in a closed system, a volatile component in equilibrium state is always present both in the liquid and the vapor phase.

This is the case postulated by the model of evaporative accretion.

7.4 Siderophile elements

In a melt containing metallic and silicate constituents, the siderophile elements (W, P, Co, Ni, Re, Os, Ir, Pt, etc.) are concentrated in a metallic phase. Silicate mantles of the differentiated celestial bodies, including the Earth and the Moon, are significantly depleted in siderophile elements because they strongly partition into the metallic cores.

There are detailed experimental data regarding distribution of the siderophile elements between silicate and metallic melts. The ratio between concentrations of a certain element in the silicate and metallic phases in equilibrium state is known as a partition coefficient: $D_{met.liq\text{-}sil.liq} = C_{met.liq}/C_{sil.liq}$.

The well-known problem of the siderophile elements in the geochemistry of the Earth is that the Earth mantle is depleted in siderophile elements to a much lesser extent than could be expected on the basis of the known partition coefficients, at least for normal temperature and pressure values (Table 7.1).

Table 7.1. Siderophile element depletion in the Earth mantle relative to the chondritic (CI) composition.

Element	Depletion in Earth mantle	
	observed*	calculated**
V	1.8	1,2
W	16.7	11,5
P	43	48
Co	13.1	105
Mo	55.6	750
Ni	14.5	1,500
Re	420	$4.8 \cdot 10^4$
Ir	330	$5.1 \cdot 10^5$

* Walter et al. (2000); Galimov (2004)
** In equilibrium with $D_{melt\text{-}sil}$ coefficients presented in Table 7.2

There are three explanations. The first is that at the final stage of the Earth's accumulation, when core segregation was completed, a portion of chondritic material fell onto the Earth's surface ("late veneer"). Its admixture enhanced the content of siderophile elements in the Earth's mantle (Wanke, 1981; Wanke et al., 1984; Yi et al., 2000; Becker et al., 2006; Albarede, 2009). However, the "late veneer' model is unable to reproduce the observed pattern of siderophile elements in the Earth's mantle (Kramers, 1998; and Tolstikhin and Kramers, 2008).

The second explanation is that fractionation of the siderophile elements between the silicate and metallic phases was 'inefficient' (i.e., there was no equilibrium). And the third explanation is that the partitioning may have been an equilibrium type, if one considers dependence of $D_{met\text{-}sil}$ on temperature, pressure and other parameters. Indeed,

Section 7.4 Siderophile elements

recent experiments have demonstrated that the partition coefficient largely depends on the P- and T-conditions, although these dependences are not simple. The siderophile behavior often shows a decrease at higher temperatures (Li and Agee, 1996; Chabot et al., 2005). But sometimes an increase is observed (Capobianco et al., 1993; Righter et al., 1997, 2009). Pressure notably affects the siderophilic property. Such elements as Ni, Co, W, P and Ga become less siderophile with increasing pressure, while Si and Mn become more siderophile as pressure increases (Li and Agee, 1996; Gessman et al., 2001; Wade and Wood, 2005). Partitioning of siderophile elements between metal and silicate melt shows a dependence on the silicate composition (Walter and Thibanlt, 1995; Eggins and O'Neil, 2002). For example, the partition coefficient for Mo increases with an increase of MgO content in the silicate melt (Righter et al., 2010). Some elements show a significant compositional effect (Nb), some are insensitive (V, Cr) to silicate melt composition (Wada et al., 2008). Not only the silicate composition but also the composition of the metallic alloy is important. For many siderophile elements the partition coefficient between iron sulfides containing metal and silicate could be lower than between iron melt and silicate (Agee et al., 1995). However, Pd is concentrated in sulfide relative to metal (Fleet et al., 1991). According to Righter and Drake (2000) the presence of carbon decreases the partitioning between metallic alloy and silicate melt for Ni, Co, and Mo, while sulfur causes an increase for the same elements. Partitioning of W is sensitive to the carbon content of the metal alloy (Cottrell et al., 2009).

As the partition coefficients significantly vary depending on conditions, one can attempt to solve an inverse problem: to identify, assuming metal-silicate equilibrium, the conditions which "match" the equilibrium partition coefficients (Mc Donough, 2003; Allegre et al., 1995). It has been shown that for Co and Ni these conditions are pressure of about 40 Gpa and temperature of $\sim 3,800$ K (Gessmann and Rubie, 1998; Chabot and Agee, 2003; Wade and Wood, 2005; Rubie et al., 2003). However, Righter et al. (2010) showed that the $D_{\text{met-sil}}$ function for Mo in the PT-diagram intersects with the $D_{\text{met-sil}}$ function for P (phosphorus) at the region which corresponds to $T = 2,400\,°C$ and $P = 22.5\,\text{GPa}$, which is different to the values found for Co and Ni. It is clear that to be acceptable the trends for all siderophile elements should "intersect" in the same region.

Kegler et al. (2008) have found that equilibrium pressure corresponding to the Earth mantle Co/Fe ratio is in a range between 32 and 37 GPa, whereas the Ni/Fe ratio requires equilibration at pressures between 45 and 50 GPa. They concluded that the Ni and Co concentrations of the Earth's mantle are most likely not the result of simple metal-silicate equilibrium.

The most difficult problem of the Earth's siderophile distribution pattern is to explain the significant excess of highly siderophile elements: Pt, Os, Pd, Ir, Ru Re in the Earth's mantle. Their concentrations are several orders higher than predicted by equilibrium coefficients. So far no plausible models have been advanced to explain this.

Thus the problem of distribution of siderophile elements in the mantle of the Earth is still far from solved.

However, we are interested here mainly in the distribution pattern of siderophile elements on the Moon. And here we encounter another problem. The pressure at the lunar core-mantle boundary is only 4 GPa. Therefore variations in partition coefficients depending on PT conditions cannot be significant in the case of the Moon.

Distribution patterns of siderophile elements on the Moon and in the Earth's mantle do not significantly differ, the Moon being slightly more depleted in siderophile elements.

The rough similarity of the concentrations of siderophile elements in the Moon and in the Earth's mantle seemed to be favorable for the model of Moon formation at the expense of material from the Earth's mantle. Ringwood and Kesson (1976, 1977) concluded that the observed similarity in siderophile elements between the Moon and the Earth's mantle, in spite of very different core sizes and pressure conditions, "implied that the Moon was derived from the Earth's mantle after the Earth's core had segregated".

The Moon's greater depletion in siderophile elements was believed to be explained by the formation of a small lunar core.

However, the proof (e.g. Canup, 2004) that the Moon originated from impactor material rather than material of the Earth's mantle created significant uncertainty regarding interpretation of the Moon siderophile element distribution pattern in terms of the giant impact. One may presume that the composition of the siderophile elements in the impactor mantle was similar to that of the Earth mantle, although the impactor was smaller and, hence, distribution of temperature and pressure was different there. Besides, it is not clear to what extent the impactor core substance with excessive content of siderophile elements could have been mixed with the impact melt thrown out to the near-Earth orbit. Given these uncertainties there is no sufficient basis for considering the issue of the siderophile elements in the Moon within the framework of the mega-impact model.

Our model suggests that the Earth and Moon originated from a common source – primitive chondritic material. The main question here is: Can the observed distribution of the siderophile elements in the Moon be reconciled with its direct segregation from the initial material of CI-composition?

At first glance it seems that the higher depletion of the Moon in siderophile elements compared to Earth is difficult to reconcile with the model of Moon and Earth formation from the same primordial material of chondritic composition, as the Earth's metallic core comprises 32 % of its mass, whereas the lunar core is no larger than 3–5 % of the mass of the Moon.

However, we will show that the problem can be solved by introducing the factor of partial melting.

Siderophile elements fractionate not only between metal and silicate, but also between solid silicate and silicate melt. Some of them (incompatible) concentrate predominantly in the liquid phase; some of them (compatible) are retained in the solid phase.

The partitioning of siderophile elements between silicate solid and silicate melt in equilibrium is described by the coefficient $D_{\text{solid-sil.liq}} = C_{\text{solid}}/C_{\text{sil.liq}}$.

I reproduce here the table (Table 7.2) from my previous work (Galimov, 2004). The estimated values of depletion of the Moon in siderophile elements used in that article almost precisely coincide with the median values for the same elements in the summary table compiled by Walter et al. (2000). However, the latter authors pointed out the wide range of uncertainties. For example, given the median value of Mo depletion in the Moon to be 1204 (Newsom (1986) followed by Galimov (2004) accepted the value 1,200) the estimates from different sources vary from 250 to 5,263 (Walter et al., 2000). In general, these uncertainties, as mentioned, are caused by variations of siderophile element fractionation depending on T, P, oxygen fugacity, compositions of silicate and metallic phases. As the pressure in the Moon interior is not extremely high one can confidently use the partition coefficient values obtained for normal PT conditions. Only a few siderophile elements are selected for consideration in Table 7.2. Vanadium is a slightly siderophile element as well as Mn and Cr. However, the latter two are also slightly volatile, which introduces an element of uncertainty to their geochemical behavior. Tungsten, phosphorus, cobalt, molybdenum and nickel are moderately siderophile elements. Among them Mo is the best for testing the model. This is a refractory element. It has pronounced siderophile properties, and simultaneously is significantly incompatible. Phosphorus is volatile and incompatible while Co and Ni are compatible. Copper, gallium, germanium, arsenic, silver, antimony, tin, and lead also show siderophile features. However, they are rather halcophile elements and their abundance is often uncertain due to volatility (Walter et al., 2000). Highly siderophile elements are represented by rhenium and iridium. They do not partition between silicate solid and silicate melt ($D_{\text{solid-sil.liq}} = 1$). Other highly siderophile elements: platinum, osmium, palladium, rhodium, gold and ruthenium in equilibrium state are almost completely concentrated in the metallic phase with $D_{\text{Met-Sil}}$ values exceeding 10^7 for Re, Pd and Au, and as high as $10^{12}-10^{15}$ for Rh, Ru, Pt. The estimated concentrations of Re (0.01ng g^{-1}), Os, Ir, Ru, Pd (0.1ng g^{-1}) and Pt (0.2ng g^{-1}) in the lunar mantle are more than 20 times lower than in the Earth's mantle (Day and Walker, 2011). Highly siderophile elements (HSE) cannot provide reliable constraints, since even a small addition of the primitive material after core segregation could change the HSE content in the mantle.

We have used the Newsom (1986) data on depletion in siderophile elements relative to carbonaceous chondrites, because they were normalized to refractory elements. And we strongly maintain the position that the Moon is enriched in refractory elements. Some authors attempt to take into account the relative contribution of material of different chondritic types (Walter et al., 2000). However, these procedures are model-dependent and may lead to prejudiced estimates.

It was mentioned above that the Moon cannot contain a metallic core larger than 5% of the total lunar mass. Thus one can take the content of the metallic melt to be at maximum about $M_{met} = 0.05$. In fact the value 0.053 has been used, as will be explained below. The calculated depletion of the Moon in siderophile elements under conditions of total melting is shown in Table 7.2.

Table 7.2. Siderophile element depletion in the Moon relative to the chondritic (CI) composition calculated for $M_{met} = 5.3\%$ and degree of partial melting $\xi = 13.2\%$.

Element	$D^{1*}_{met/silicate}$	$D^{2*}_{solid/liquid}$	Depletion ($1/f$)		
			Observed [1*]	Calculated	
				total melting	partial melting
V	1.7	0.004	1.9	1.03	1.71
W	36	0.01	22	2.9	22.0
P	160	0.05	115	9.4	67.6
Co	350	1.8	12	19.5	11.6
Mo	2,500	0.05	1200	133.4	1042
Ni	5,000	5	50	266	57.6
Re	$1.6 \cdot 10^5$	1	$3 \cdot 10^4$	$8.5 \cdot 10^3$	$8.5 \cdot 10^3$
Ir	$1.7 \cdot 10^6$	1	$0.9 \cdot 10^5$	$0.9 \cdot 10^5$	$0.9 \cdot 10^5$

[1*] The data mostly from Newsom (1986) and references therein.
[2*] Drake (1986). There are more recent estimates for the abundances of siderophile elements and the partition coefficients but the values used in this study are within the uncertainty limits of the new values. Therefore we retain Table 2 as it was presented in Galimov (2004).

Looking at Table 7.2 one can see an interesting correlation. Incompatible ($D_{sol-liq} < 1$) elements (V, W, P, Mo) deplete the lunar material to a higher degree than that calculated for the case of its total melting, while the compatible elements (Co, Ni) overburden the lunar material.

This means that partial melting and partitioning of siderophile elements between solid and liquid phases may play a role. In other words the degree of Moon depletion in siderophile elements is determined not only by partition in the metal-silicate system but also in the solid-liquid system (Newsom, 1986). It should be stressed that a particularity of our model is that metal is not in exchange with the whole silicate volume, but only with the melted part. We have found that the formulas in the Newsom (1986) work do not reproduce this situation. Therefore the inference of the equation used is shown below.

The expression for the depletion factor (f) of matter in a siderophile element accounting for both types of fractionation, due to partial melting and partitioning in the metal-silicate system, can be easily obtained from the following considerations.

In accordance with the model shown in Fig. 7.5, equilibrium distribution of the siderophile element takes place between the solid phase and the silicate liquid with

Section 7.4 Siderophile elements

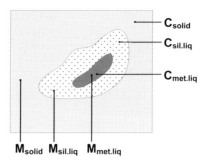

Figure 7.5. On the discussed model of distribution of siderophile elements in partial melting. Mass of phases respectively: solid state – M_{solid}, silicate melt – $M_{sil.liq}$, metallic melt – $M_{met.liq}$ concentrations of a siderophile element respectively: in silicate solid – C_{solid}, in silicate melt – $C_{sil.liq}$, in metallic melt – $C_{met.liq}$.

the $D_{solid\text{-}sil.liq}$ ratio, and also between the metallic melt and the silicate melt with the $D_{met.lig\text{-}sil.liq}$ ratio.

The ratio between the phases is described by the following parameters: ξ – partial melting degree (mass fraction of the melt): $\xi = \frac{M_{liq}}{M_o}$, where M_{liq} – mass fraction of the melt including the silicate and metallic components, and M_o is the summarized mass of a volume unit that includes the solid phase.

$$\eta = \frac{M_{met.liq}}{M_o} \quad \text{where } \eta \text{ is the mass fraction of the molten metal.}$$

In this case, the following ratios exist:

1. $M_o = M_{solid} + M_{sil.liq} + M_{met}$,
2. $M_{liq} = M_{sil.liq} + M_{met}$,
3. $M_o C_o = M_{solid} \cdot C_{solid} + M_{sil.liq} \cdot C_{sil.liq} + M_{met} \cdot C_{met}$,
4. $D_{solid\text{-}sil.liq} = \dfrac{C_{solid}}{C_{sil.liq}}$,
5. $D_{met\text{-}sil.liq} = \dfrac{C_{met}}{C_{sil.liq}}$,
6. $\xi = \dfrac{M_{liq}}{M_o}$,
7. $\eta = \dfrac{M_{met}}{M_o}$,
8. $f = \dfrac{M_{solid} \cdot C_{solid} + M_{sil.liq} \cdot C_{sil.liq}}{C_o(M_{solid} + M_{sil.liq})}$,

where f – degree of depletion of the silicate phase in a siderophile element due to its move to the metallic phase.

Let us perform certain transformations:

$$f = \frac{M_o C_o - M_{\text{met}} \cdot C_{\text{met.liq}}}{C_o (M_o - M_{\text{met.liq}})} \tag{7.1}$$

The numerator and the denominator are divided by $M_o C_o$

$$f = \frac{1 - \frac{M_{\text{met}} \cdot C_{\text{met.liq}}}{M_o \cdot C_o}}{1 - \frac{M_{\text{met}}}{M_o}} = \frac{1 - \eta \frac{C_{\text{met}}}{C_o}}{1 - \eta} = \frac{1 - \eta \frac{C_{\text{met}}}{C_o} \cdot \frac{C_{\text{sil.liq}}}{C_{\text{sil.liq}}}}{1 - \eta}$$

$$= \frac{1 - \eta \cdot D_{\text{met-sil.liq}} \cdot \frac{C_{\text{sil.liq}}}{C_o}}{1 - \eta}. \tag{7.2}$$

The expression (3) is transformed as follows:

$$1 = \frac{M_{\text{solid}}}{M_o} \cdot \frac{C_{\text{solid}}}{C_o} + \frac{M_{\text{sil.liq}}}{M_o} \cdot \frac{C_{\text{sil.liq}}}{C_o} + \frac{M_{\text{met}}}{M_o} \cdot \frac{C_{\text{met}}}{C_o}.$$

Both parts are multiplied by $\frac{C_o}{C_{\text{sil.liq}}}$

$$\frac{C_o}{C_{\text{sil.liq}}} = \frac{M_{\text{solid}}}{M_o} \cdot \frac{C_{\text{solid}}}{C_{\text{sil.liq}}} + \frac{M_{\text{sil.liq}}}{M_o} + \frac{M_{\text{met}}}{M_o} \cdot \frac{C_{\text{met}}}{C_{\text{sil.liq}}}$$

$$= \frac{M_o - M_{\text{liq}}}{M_o} \cdot D_{\text{solid-sil.liq}} + \frac{M_{\text{liq}} - M_{\text{met}}}{M_o} + \eta \frac{C_{\text{met}}}{C_{\text{sil.liq}}}$$

$$= (1 - \xi) D_{\text{solid-sil.liq}} + \xi - \eta + \eta \cdot D_{\text{met-sil.liq}}. \tag{7.3}$$

The joint solution of these equations results in the following expression for f:

$$f = \frac{\xi - \eta + (1 - \xi) D_{\text{solid-sil.liq}}}{(1 - \eta)[(1 - \xi) D_{\text{solid-sil.liq}} + \xi - \eta + \eta D_{\text{met-sil.liq}}]}. \tag{7.4}$$

This formula (subject to modifications of certain designations) coincides with the corresponding expression derived by Galimov (2004).

In the case of complete melting ($\xi = 1$) as well as in the case this element does not split into solid silicates and the melt ($D_{\text{solid-sil.liq}} = 1$):

$$f = \frac{1}{1 - \eta(1 - D_{\text{met-sil}})}. \tag{7.5}$$

We proceed further in the following manner. Among the siderophile elements, Ir is practically not fractionated between the solid phase and melt ($D^{\text{Ir}}_{\text{solid-sil.liq}} = 1$). For the equilibrium partitioning of Ir between silicate and metal (i.e., controlled by the respective coefficient $D^{Ir}_{\text{met-sil.liq}}$) and the observed iridium depletion on the Moon ($1/f = 0.9 \times 10^5$), this expression yields $\eta = 0.053$. This estimate can be considered as the maximum relative mass of the lunar metal (5.3 %) corresponding to its efficient (equilibrium) segregation from the primordial lunar material. In fact not all the iron that participated in the fractionation of siderophile elements could be accumulated in the core.

Given the obtained η_{met}, it is possible to estimate the degrees of depletion of various siderophile elements in the Moon for different ξ values. Table 7.2 shows the result of such calculations for $\xi = 13{,}2\%$. As can be seen from the table, in the case of complete melting ($\xi = 1$), there is no agreement between the calculated and the observed parameters. On the other hand, the observed distribution patterns of siderophile elements on the Moon are adequately approximated by the values calculated assuming that iron was segregated under conditions of partial melting. The optimum value of the degree of melting is $\xi = 13, 2\%$.

For example, for Mo, in the case of partial melting: ($D^{Mo}_{\text{solid-sil.liq}} = 0.05\xi = 13\%$ and $\eta = 0.053$, $D^{Mo}_{\text{met-sil.liq}} = 2,500$) according to Equation (7.4) we get $\frac{1}{f} = 1,042$, i.e., the depletion is 1,042-fold. If the metal content is the same (5.3 %), in the case of complete melting we will get according to Equation (7.5): $f = \frac{1}{133.4}$, i.e., the depletion is 133.4-fold. The observed depletion in phosphorus is somewhat higher than the model value. But unlike other elements listed in Table 7.2, phosphorus is volatile. Its elevated depletion is therefore explained by this additional factor.

Thus the concept of the Moon formation from a primary material of the CI composition can be reconciled with the observed abundances of siderophile elements.

We do not link the segregation process of the metals and the siderophile elements on the Moon with any specific geological scheme. It may have been exposure of metal from magmatic melts, which is in line with the concept of the allegedly existing lunar magmatic ocean; after exposure the deposited metal was equilibrated with the KREEP-type layer containing incompatible elements. It may also have been percolation of metal via rocks that are partially melted. The last case is, in principle, relevant to the Moon conditions, since the lunar substance consists of a more refractory material than that of the Earth.

The question arises whether the segregation of metal is possible under such low degrees of melting. We have tried to study this experimentally (Galimov, Lebedev, 2012). The process of metal-silicate segregation has been studied in numerous works (Ballhaus et al., 1996; Shannon and Agee, 1998; Roberts et al., 2007; Lebedev et al., 1999; Bagdassarov et al., 2009; Holzheid et al., 2000; Jana and Walker, 1996; etc.). As iron shows a low wettability with respect to silicates the segregation and percolation of iron through a silicate rock is believed to require a significant degree of melting. The presence of sulfide iron increases wetting. It should also be kept in mind that the lunar material is enriched in refractory oxides such as Al_2O_3, CaO and TiO_2. Therefore the solidus temperature of the lunar mantle must be higher, and easier permeable by a high-temperature iron melt. In addition it has been demonstrated that the permeability increases significantly if the silicate-metal system is subjected to mechanical deformations (T. Rushmer, 1995; T. Rushmer et al., 2000). In the above-mentioned experiment (Lebedev and Galimov, 2012) we used a technique of high temperature centrifugation (Kadik and Lebedev, 1989). The starting composition consisted of: the metal-sulfide phase $\sim 5\%$ (95Fe : 5S) as 1–5 μk powder dispersed in silicate; solid silicate matrix $\sim 85\%$ (olivine crystals on 89 % of forsterite); and 10 % of ferrobasalt and ferropikrite which formed the intergrain melt. The deforming tensions were simulated by use of high density plummet. The oxygen fugacity was 4–5 units below IW. Centrifugation during 15 minutes at $T = 1,440\,°C$ and $G = 4,000\,g$ led to accumulation of metallic phase at ~ 1.5 mm of the lower part of the 10 mm height ampule. The metallic phase was presented by about $\sim 100\,\mu$k blobs (Fig. 7.6).

Thus segregation of the metallic core in the Moon under the condition of a relatively low degree of partial melting evidently may occur.

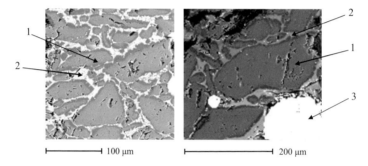

Figure 7.6. Experimental study of metal-silicate segregation by centrifugation (Lebedev and Galimov, 2012). 1 – Solid olivine crystals, 2 – ferrobasaltic melt, 3 – accumulated blobs of the metal phase.

The partition ratios we have used in our calculations are not adjusted according to their dependence on pressure, temperature and oxygen fugacity. Nevertheless, calculated values closely approximate the observed lunar distribution of the siderophile elements as compared to the CI chondrites. One should note that the recent attempt to use parameterized values of partition ratios (i.e., taking into consideration their dependence on P, T and f_{o2}) has also yielded results that conform (although not completely) with the concept that the lunar substance originated from a CI bearing material (Rao and van Westerenen, 2012).

In any case, we see that as the lunar substance is differentiated, it is possible to understand certain mechanisms leading to the observed distribution of the siderophile elements in the Moon in accordance with the composition of the CI chondrites.

7.5 Constraints following from Hf–W systematic

Hf–W systematics is most appropriate for the analysis of the problem of the formation of the Earth and the Moon.

It is known that ^{182}Hf is a short-lived isotope transforming into ^{182}W with a half-life of 8.9 Ma. Hafnium and tungsten are fractionated during iron separation from silicates. Hafnium is a lithophile element and is retained in silicates, whereas W shows a siderophile affinity and partitions into the metallic phase. Since the precursor of ^{182}W, the ^{182}Hf radioisotope, remains in the silicate part, the ^{182}W/^{184}W ratio of the silicate increases compared with undifferentiated chondritic material. The earlier the differentiation, the higher the isotopic shift (ε_w). Differentiation must occur within the earliest tens of millions of years, before the complete decay of short living ^{182}Hf. Early studies (Lee and Halliday, 1995; Lee et al., 1997; Halliday and Lee, 1999) reported rather high positive ε_w values for the Moon and negligible isotopic shifts for the Earth.

However, the analytical method used in these studies was subsequently challenged (Schoenberg et al., 2002; Yin et al., 2002). The revision of isotopic data resulted in the reestimation of the time of the formation of the Earth's core and the Moon (Yin et al., 2002). Additional uncertainty was associated with the fact that the interaction of Ta with cosmogenic neutrons, ^{181}Ta (n) → Ta $^{182}(\beta)$ →^{182}W, can significantly contribute to the production of excess ^{182}W (Leya et al., 2000). Recent investigations of W isotopic systematics in Ta-free lunar metals showed that there is no excess ^{182}W in such a case (Kleine et al., 2005; Touboul et al., 2007).

Eventually, a consensus was reached after revision of the previous results, and the following conclusions were formulated (Kleine et al., 2009).

The Hf/W ratios of the present-day Earth and Moon are different: 17 ± 5 and 26 ± 2, respectively.

The Earth and the Moon have almost identical W isotopic compositions,

$$\Delta \varepsilon^{182} W = 0.09 \pm 0.1.$$

The W isotopic compositions of the Earth and the Moon are slightly different from the chondritic composition,

$$\varepsilon^{182} W = +1.9.$$

Let us consider the consequences of these results in the context of the hypothesis of giant impact, on the one hand, and the model of evaporative accretion, on the other hand.

Since the isotopic compositions of the Moon and the Earth are identical but their Hf/W ratios are different, the observed difference had to appear after the almost complete decay of the precursor of radiogenic ^{182}W, i.e., ^{182}Hf.

Kleine et al. (2009) reported the following estimate. In order to obtain a $\Delta \varepsilon^{182}$W value within 0.09 ± 0.1 (if the value +0.09 is significant!), the age of the event must be not earlier than ∼ 50 Ma after solar system formation.

This value is an important time constraint on events in both hypotheses. In the giant impact hypothesis, this value corresponds to the time of the impact. By this moment proto-Earth and the impactor already existed as differentiated bodies. They experienced different geochemical histories before the collision. Therefore, either they accidentally had identical W isotopic characteristics, or the giant impact must have lead to the isotopic homogenization of the whole mass formed during the event. This problem is similar to that arising in the giant impact hypothesis in connection with the observed coincidence of the oxygen isotope fractionation trends of the Moon and the Earth.

The Hf–W isotope data are very difficult to explain in the framework of the giant impact model. The advocates of this hypothesis are aware of this difficulty. It is pertinent to cite Kleine et al. (2009): "that two such different objects as the proto-Earth and the impactor would evolve to identical W isotope compositions in their mantles seems highly unlikely...," and "the identical W isotope compositions of the lunar and terrestrial mantles could indicate that the Moon is largely derived from terrestrial mantle material, but this is inconsistent with results from numerical simulations, all of which indicate that the Moon predominantly consists of impactor material".

In contrast to their disagreement with giant impact model, the identical W isotopic characteristics of the Moon and Earth appear natural for the evaporative accretion model. This corresponds to the time when the geochemical histories of the Moon and the Earth diverged. Until that moment, they consisted of the same material, and their W isotopic compositions were obviously identical. The change in Hf–W composition and eventually the observed difference in Hf–W composition of the Earth and the Moon is obviously related to core formation. It means that core formation in both Earth and Moon could not have begun before 50–60 million years after the beginning of the solar system. The question remains whether fragmentation occurred at the same time, or fragmentation happened earlier but core formation started with some delay. After $\sim$ 50 Ma, the Hf/W ratios of the Earth and the Moon became different, but the W isotopic compositions could not have already changed.

The ε_w shift relative to the chondritic value both in the terrestrial and lunar materials can merely be the result of the difference between the assumed chondritic and primary proto-terrestrial (lunar) Hf/W ratios. Calculations show that the observed $\varepsilon_w = +1.9$ could be reached if the Hf/W ratio of the proto-terrestrial gas-dust body was on average 2.2 times higher than the chondritic value by 50–60 Ma. Note that the Hf/W ratio of carbonaceous chondrites, $^{180}Hf/^{184}W = 1.21 \pm 0.06$, is twice as high as that of H chondrites, $^{180}Hf/^{184}W = 0.63 \pm 0.20$ (Kleine et al., 2008).

The experimentally measured age of the oldest lunar rocks constrains any hypothesis. The upper boundary is defined by the age of the oldest rocks.

It can be seen from Table 3.2 (Chapter 3) that the age of the oldest known lunar rocks is 4.44–4.50 Ga; i.e., the first solid rocks appeared in the Moon between 50 and 70 Ma after solar system formation.

Thus the lunar core (as well as the Earth core in our model) cannot have emerged before 50 Ma after the beginning of the solar system. Fragmentation itself may have occurred earlier, if we assume that the core started to form not at the moment of the planetary body formation but with some delay.

The principal question of the evaporative accretion model is how long the accomplishment of accretion took. By the end of the accretion about 30% of mass would be added to the lunar embryo, and the Earth embryo would enlarge by 20 to 30 times, acquiring the dominant part of the initial vapor-particle formation. In an attempt to answer this question we turn to consideration of U–Pb and Rb–Sr systematic. But before that, a few words about the Xe-isotopes system.

7.6 ^{129}I–^{129}Xe and ^{244}Pu–^{136}Xe

There is a problem in the isotopic geochemistry of xenon, which is sometimes referred to as the xenon paradox (Podosek and Ozima, 2000; Swindle and Podosek, 1988; Allegre et al., 1995). The Xe-isotope systematic is one of the most complicated and enigmatic in existence (Reyholds, 1963; Ozima, 1994; Pepin and Porcelly, 2006).

Paradoxically, the isotopic composition of terrestrial Xe indicates that the Earth lost the Xe that was formed during the first 110–130 Ma, although it should have been retained by gravity.

The fractions of radiogenic ^{129}Xe and ^{136}Xe, which are produced by the decay of ^{129}I and ^{244}Pu, respectively, in atmospheric Xe are much lower than could be expected (Podosek and Ozima, 2000). The relatively short-lived ^{129}I and ^{244}Pu isotopes have half-lives of 17 and 81 Ma, respectively. The observed proportion of Xe isotopes in the Earth atmosphere is such that the Earth missed Xe of the short living ^{129}I–^{129}Xe and ^{244}Pu–^{136}Xe systems, and retained xenon only after 110–130 Ma after solar system formation (Podosek and Ozima, 2000; Allegre et al., 2008). Yet, an anomalous ^{129}Xe occurs in terrestrial rocks, that was probably generated by spontaneous fission of ^{238}U and neutron-induced fission of ^{235}U (Shukolukov and Meshik, 1994; Pinti et al., 2001), there is a significant deficit of ^{129}Xe in the terrestrial atmosphere.

Leaving aside some exotic hypotheses, the explanation of this fact is reduced to the suggestion that the Earth's primordial atmosphere was lost in a catastrophic event at $\sim$ 110–130 Ma (Porcelli et al., 2001).

According to Halliday (2008), this event was the giant impact that produced the Moon. But lunar rocks exist that formed earlier than 100 Ma after solar system formation (Table 3.2). If the Moon-forming giant impact occurred earlier, this excludes a subsequent large impact that could have resulted in the loss of the atmosphere (Canup and Asphaug, 2001).

Within our concept the loss of the early Xe is a consequence of volatile removal in a hydrodynamic flux during the compression of the proto-planetary vapor-particle cloud. The model interpretation suggests that by 110–130 Ma hydrodynamic escape ceased and the Earth began to retain xenon. Obviously, this is the time when the accretion process was terminated.

Note that, as is known, terrestrial Xe is fractionated relative to meteoritic Xe (Podosek and Ozima, 2000). Such fractionation is possibly due to gravitational separation during hydrodynamic escape (Hunten et al., 1987). This effect is incompatible with instantaneous loss of the atmosphere.

7.7 U–Pb system. Time of accomplishment of accretion

Similar to Xe, more than 98% of initial terrestrial Pb was lost. Primary (solar) Pb shows a ^{238}U/^{204}Pb ratio (designated as μ) of 0.27 (Anders and Grevesse, 1989), whereas terrestrial Pb has a $^{238}\mu$ of 8–10 (Allegre et al., 1995).

The lunar ^{238}U/^{204}Pb ratio (μ_L-value) is estimated to be 27 ± 30 (KREEP-sample 78,238, Edmunson et al., 2008), ~ 35 (sample 60,025 by Premo and Tatsumoto, 1992), 70 ± 30 (mare basalt sample 10,017, Gaffney et al., 2007), and as high as ~ 508 (sample 78,236 by Premo and Tarsumoto, 1991).

Thus the Moon lost much more lead than the Earth, but quantitatively this depletion is poorly constrained.

The present day Pb isotope ratios for the Earth are: $\left(\frac{^{206}\text{Pb}}{^{204}\text{Pb}}\right)_{present} = 18.279$ and $\left(\frac{^{207}\text{Pb}}{^{204}\text{Pb}}\right)_{present} = 15.491$ while for the Moon these ratios are estimated in limits 160–190 and 80–340 respectively.

Allegre et al. (2008) suggested that the loss of early Earth's Xe and the loss of early Pb were related and reflected a common major event during Earth's differentiation at ~ 4.45 Ga (~ 117 Ma after solar system formation). In the case of Pb, its removal into the core in a sulfide form can be considered. It is supposed that the core segregation that occurred at the time was accompanied by a tremendous energy release resulting in the formation of the magma ocean and the loss of the atmosphere.

The suggestion that Pb was removed to the core does not explain why the Moon is more depleted in Pb than the Earth, although its core is much smaller. Lead is in fact a lithophile rather than siderophile element, although it can show siderophile properties in a sulfide form at low redox potential. But its affinity to the metal phase is hardly sufficient to explain more than 60-fold depletion in the Earth's mantle, higher than that of some typical siderophile elements, for instance, W and Co.

It is a reasonable suggestion that Pb has escaped, similar to Xe, by being carried away by a hydrodynamic flux. Lead is one of the most volatile elements. Its depletion in the Earth and, to a greater extent, in the Moon is logically consistent with our model.

Let us assume that hydrodynamic escape occurred in a period from $t = 0$, and ceased at $t = t_*$, when the loss of Pb terminated. Then for the period from t_* to $t_{present} = 4.568$ when the content of ^{204}Pb remained constant, the following relation holds for the ^{238}U-^{206}Pb system:

$$\left(\frac{^{206}\text{Pb}}{^{204}\text{Pb}}\right)_{present} = \left(\frac{^{206}\text{Pb}}{^{204}\text{Pb}}\right)_* + {}^{238}\mu[e^{\lambda_{238}(4.568-t_*)} - 1], \qquad (7.6)$$

Section 7.7 U–Pb system. Time of accomplishment of accretion

where $(^{206}Pb/^{204}Pb)_*$ is the isotopic composition of Pb at the moment of termination of Pb loss.

For the period from $t = 0$ to $t = t_*$ the following relation would have held, if ^{204}Pb content were constant:

$$\left(\frac{^{206}Pb}{^{204}Pb}\right)_* = \left(\frac{^{206}Pb}{^{204}Pb}\right)_0 + {}^{238}\mu \cdot e^{\lambda_{238}(4.568-t_*)}(e^{\lambda_{238}t_*} - 1). \tag{7.7}$$

But, if we have a situation of ^{204}Pb change during this period, then $^{238}\mu$ should be substituted by some effective value, μ_{eff}. Let term $e^{\lambda}_{238}{}^{(4.568-t_*)}$ be denoted as a_{238}. Then,

$$\left(\frac{^{206}Pb}{^{204}Pb}\right)_* = \left(\frac{^{206}Pb}{^{204}Pb}\right)_0 + \mu_{\text{eff}} \cdot a_{238}(e^{\lambda_{238}t_*} - 1). \tag{7.8}$$

The factor a_{238} denotes the higher content of ^{238}U at $t = t_*$ compared to the present time, related to the radioactive decay of ^{238}U.

The term which stands for $\mu_{\text{eff}} \cdot a_{238}$ in Equation (7.8) is variable depending on the current content of ^{204}Pb as a function $\varphi_{(U/Pb)_t}$ in the time scale from $t = 0$ to $t = t_*$. In Equation (7.8) it may be expressed as an integrated mean value of function $\varphi_{(U/Pb)_t}$:

$$\mu_{\text{eff}} \cdot a_{238} = \frac{1}{t_*} \int_0^{t_*} \varphi_{238}(t) dt. \tag{7.9}$$

The limits of the integral for the $\mu_{\text{eff}} \cdot a_{238}$ value are from $(^{238}U/^{204}Pb)_0 = 0.27$ at $t = 0$ to $^{238}\mu \cdot a_{238}$ at $t = t_*$.

Similarly, for the $^{235}U-^{207}Pb$ system:

$$\left(\frac{^{207}Pb}{^{204}Pb}\right)_{\text{present}} = \left(\frac{^{207}Pb}{^{204}Pb}\right)_* + \frac{^{238}\mu}{137.88}[e^{\lambda_{235}(4.568-t_*)} - 1], \tag{7.10}$$

$$\left(\frac{^{207}Pb}{^{204}Pb}\right)_* = \left(\frac{^{207}Pb}{^{204}Pb}\right)_0 + \frac{\mu_{\text{eff}}}{137.88}e^{\lambda_{235}(4.568-t_*)}(e^{\lambda_{235} \cdot t_*} - 1). \tag{7.11}$$

Factor $a_{235} = e^{\lambda}_{235}(4.568 - t_*)$, and

$$\frac{1}{137.88}\mu_{\text{eff}} \cdot a_{235} = \frac{1}{t_*} \int_0^{t_*} \varphi_{235}(t) dt. \tag{7.12}$$

As the magnitude of μ_{eff} depends almost entirely on ^{204}Pb variations it has the same meaning and value in both $^{238}U-^{206}Pb$ and $^{235}U-^{207}Pb$ systems. The change of ^{238}U and ^{235}U due to decay during t_0 to t_* influences insignificantly (by a few percentage points) the values $\mu_{\text{eff}} \cdot a_{238}$ and $\mu_{\text{eff}} \cdot a_{235}$. This might be considered by introducing an expression of the type $^{238}U(t)/^{238}U_0 = e^{-\lambda t}$ into the integral expression.

But it is unreasonable from the standpoint of simplicity bearing in mind comparable uncertainties of other parameters.

Expressing $(^{206}\text{Pb}/^{204}\text{Pb})_*$ in Equations (7.6)–(7.8) and $(^{207}\text{Pb}/^{204}\text{Pb})_*$ in Equations (7.10)–(7.11), and using the known values of $\lambda_{238} = 0.155125 \text{ Gyr}^{-1}$, $\lambda_{235} = 0.98485 \text{ Gyr}^{-1}$, $(^{206}\text{Pb}/^{204}\text{Pb})_{\text{present}} = 18.279$, $(^{207}\text{Pb}/^{204}\text{Pb})_{\text{present}} = 15.491$, and initial solar ratios (Canion Diablo meteorite) $(^{206}\text{Pb}/^{204}\text{Pb})_0 = 9.307$ and $(^{207}\text{Pb}/^{204}\text{Pb})_0 = 10.294$, we obtain:

$$18.279 = 9.307 + \mu_{\text{eff}} e^{0.155125(4.568-t_*)}(e^{0.155125 t_*} - 1)$$
$$+ {}^{238}\mu [e^{0.155125(4.568-t_*)} - 1], \tag{7.13}$$

$$15.491 = 10.294 + \frac{\mu_{\text{eff}}}{137.88} e^{0.98485(4.568-t_*)} \cdot (e^{0.98485 t_*} - 1)$$
$$+ \frac{{}^{238}\mu}{137.88}[e^{0.98485(4.568-t_*)} - 1]. \tag{7.14}$$

Joint solution of Equations (7.13) and (7.14) determines μ_{eff} and t_*.

Given ${}^{238}\mu = 9$, we obtain $t_* = 120$ Ma, $\mu_{\text{eff}} = 0.7$, $a_{238} = 1.99$, and $a_{235} = 79.887$. Correspondingly, the mean values of functions are $(\varphi_{238})_{\text{mean}} = \mu_{\text{eff}} a_{238} = 1.4$ and $(\varphi_{235})_{\text{mean}} = \mu_{\text{eff}} a_{235} \cdot 1/137.88 = 0.41$.

At the time t_* the $(^{238}\text{U}/^{204}\text{Pb})_*$ ratio is ${}^{238}\mu a_{238} = 9 \cdot 1.99 = 17.9$, and the $(^{235}\text{U}/^{204}\text{Pb})_*$ ratio is $1/137.88 \cdot {}^{238}\mu a_{235} = 9 \cdot 1/137.88 \cdot 79.887 = 5.21$. Thus during the first 120 Ma the ${}^{238}\text{U}/^{204}\text{Pb}$ ratio changes from value 0.27 to 17.9, and the ${}^{235}\text{U}/^{204}\text{Pb}$ ratio changes from 0.085 to 5.21. There is a small difference in ${}^{204}\text{Pb}$ depletion coefficients in these systems: $17.9 : 0.27 = 66.3$ and $5.21 : 0.085 = 61.3$. The coefficients may be corrected by taking into account contributions from U decay. According to radioactive decay low ${}^{238}\text{U}_{(t_*=0.12)} = {}^{238}\text{U}_0 e^{-\lambda_{238} 0.12}$ the correcting factors are 0.98 and 0.89 for ${}^{238}\text{U}$ and ${}^{235}\text{U}$ respectively. It yields the consistent value of about 68 for ${}^{204}\text{Pb}$ depletion in both systems. Note that it is higher than the 30-fold depletion usually mentioned in literature that is inferred from comparison of the present ${}^{238}\text{U}/^{204}\text{Pb} = {}^{238}\mu = 9$ to the initial ratio 0.27.

Since the escaping component of the U–Pb system appears in the denominator it is reasonable to use a reciprocal function $\varphi_{238}^{-1}(t)$ and $\varphi_{235}^{-1}(t)$.

Correspondingly, the scale of values for ${}^{204}\text{Pb}/^{238}\text{U}$ is the following:

$$(^{238}\text{U}/^{204}\text{Pb})_0^{-1} = 0.27^{-1} = 3.7, (\varphi_{238})_{\text{mean}}^{-1} = 0.71, (^{238}\mu a_{238})^{-1} = 0.06;$$

and for

$$^{204}\text{Pb}/^{235}\text{U} : (^{235}\text{U}/^{204}\text{Pb})^{-1} = 0.085^{-1} = 11.72; (\varphi_{238})_{\text{mean}}^{-1} = 2.43$$

and $(^{235}\mu a_{235})^{-1} = 0.19$.

The $(\varphi_{238})_{\text{mean}}^{-1}$ and $(\varphi_{235})_{\text{mean}}^{-1}$ curves should be identical in the respective scales.

The exact form of the curves is unknown. However, their approximate shape can be deduced from the rule of plotting of an averaging function: the areas in a plot contained between the average line and the curve above and the curve below must be equal. In Fig. 7.7 an exponential curve is selected to approximate the evolution of the Pb/U ratio under the above parameters.

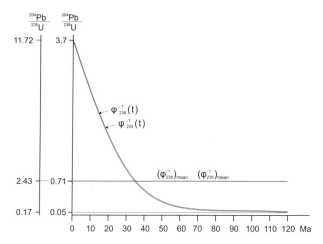

Figure 7.7. Evolution of ^{204}Pb/U ratio during first 120 Ma of the solar system history due to Pb hydrodynamic escape suggested by the evaporative accretion model.

Thus the analysis of the U–Pb system allows interpretation of the observed deficit of the primordial ^{204}Pb in the Earth's mantle within the framework of the suggested model as a result of hydrodynamic escape during the first ~ 120 Ma of the solar system history.

Great uncertainty in recent times about the bulk lunar ^{238}U/^{204}Pb and ^{235}U/^{204}Pb ratios (see the beginning of the section) does not allow applying U–Pb systematic analysis directly to the Moon. The Rb-Sr systematic is more applicable here.

7.8   Rb-Sr system. Time of Moon origin

Rubidium-87 is a long-lived isotope with a half-life of $T_{1/2} = 48.8 \cdot 10^9$ yr and a decay constant of $\lambda_{87} = 1.42 \cdot 10^{-11}$ yr^{-1}.

The starting point of the evolution of the Rb-Sr system is defined by the solar ^{87}Rb/^{86}Sr= 0.92 (Grevese et al., 1998) and the initial Sr isotope ratio of the earliest CAI: $(^{87}$Sr/^{86}Sr$)_0 = 0.69892$ (Carlson and Lugmair, 1988), and the initial Sr isotope ratio $(^{87}$Sr/^{86}Sr$)_{i\text{-Moon}} = 0.69906$ (Carlson and Lugmair, 1988).

Both the Earth and the Moon are significantly depleted in Rb. The Earth has ^{87}Rb/^{86}Sr $= 0.09$, and the Moon has ^{87}Rb/^{86}Sr $= 0.018$.

Rb is a volatile element, but less volatile than Pb (Fig. 6.4, Chapter 4). Its volatility is comparable with the volatility of other alkali elements (Na, K, Cs). The Earth is depleted in Rb 10-fold (0.92:0.09), while it is depleted in Pb 68-fold as has been shown in the previous section. As in the case of Pb it is reasonable to suggest that loss of Rb occurred due to hydrodynamic escape during contrasstion of the proto-Earth-Moon vapor-particle formation.

Similar to the ^{238}U/^{204}Pb and ^{235}U/^{204}Pb systems the Sr isotope composition in a given moment t_* is specified as

$$\left(\frac{^{87}Sr}{^{86}Sr}\right)_* = \left(\frac{^{87}Sr}{^{86}Sr}\right)_o + \frac{^{87}Rb}{^{86}Sr}[e^{\lambda_{87}4.568} - e^{\lambda_{87}(T_o-t_*)}]. \tag{7.15}$$

If the Rb/Sr ratio changes owing to Rb loss, the structure of Equation (7.15) can be retained by introducing the effective ratio (Rb/Sr)$_{eff}$:

$$\left(\frac{^{87}Sr}{^{86}Sr}\right)_* = \left(\frac{^{87}Sr}{^{86}Sr}\right)_o + \left(\frac{^{87}Rb}{^{86}Sr}\right)_{eff} \cdot e^{\lambda_{87}(4.568-t)}(e^{\lambda_{87} \cdot t_*} - 1). \tag{7.16}$$

The exponent term $e^{\lambda_{87}(4.568-t_*)}$ is denoted as a_{87}.

As the half-life of ^{87}Rb is very high the a_{87} is not susceptible to the exact t_*. For the first few tens of millions of years the a_{87} value is close to 1.07.

$$[\varphi_{Rb}(t)]_{av} = 1.07\left(\frac{^{87}Rb}{^{86}Sr}\right)_{eff} = \frac{1}{t_*}\int_0^{t_*} \varphi_{Rb}(t) \cdot dt, \tag{7.17}$$

where $\varphi_{Rb}(t)$ is the law of ^{87}Rb/^{86}Sr alteration in the initial gas-dust formation. The integral limits are known. These are 0.92 at $t = 0$ and 0.09 at $t = t_*$.

If we assume $t_* = \sim 120$ Ma (from U–Pb system analysis) and know $(^{87}Sr/^{86}Sr)_{t_*}$, i.e., the initial Sr isotope composition of Earth $(^{87}Sr/^{86}Sr)_{i\text{-Earth}}$, we will be able to find $(^{87}Rb/^{86}Sr)_{eff}$, and then, as the $(^{87}Sr/^{86}Sr)_{i\text{-Moon}}$ is exactly known, to obtain the time of the Moon formation in our model. Unfortunately $(^{87}Sr/^{86}Sr)_{i\text{-Earth}}$ is poorly determined. Unlike the U–Pb systematic where we have two isotopic systems, here we have only one equation for two variables. Therefore we have to make a reasonable assumption.

The Moon, like achondrites, has a sufficiently low Rb/Sr ratio to enable its initial Sr-isotope composition to be determined with adequate accuracy. As mentioned $(^{87}Sr/^{86}Sr)_{i\text{-Moon}} = 0.69906$. For the Earth $(^{87}Sr/^{86}Sr)_{i\text{-Earth}}$ cannot be precisely determined. It was conventionally accepted for a long time that the Earth's Sr-isotope was initially identical to that of achondrites. The latter is determined as BABI = 0.69897 (Basaltic Achondrites Best Initial, Papanastassiou and Wasserburg, 1969). However, subsequent studies have shown the inadequacy of this suggestion. H. Palme (2001) suggested $(^{87}Sr/^{86}Sr)_{i\text{-Earth}}$ to be 0.69902, but with a significant uncertainty of $\pm 8 \cdot 10^{-5}$. Kostitsyn (2004) showed that the initial terrestrial isotopic ratios of Sm/Nd, Lu/Hf, and Rb/Sr contradict the idea of their being identical with the initial chondritic reservoir (CHUR). Recently I. Campbell and H. O'Neil (2012) have presented evidence against a chondritic Earth. At present the value $(^{87}Sr/^{86}Sr)_{i\text{-Earth}} = 0.69940$ suggested by McCulloch (1994) seems to be the most tenable.

Section 7.8 Rb-Sr system. Time of Moon origin

To solve Equation (7.16) adopting McCulloch's value, then:

$$0.69940 = 0.69892 + (^{87}\text{Rb}/^{86}\text{Sr})_{\text{eff}} a_{87}(e^{1.42 \cdot 10^{-11} \cdot 0.12 \cdot 10^{-9}} - 1)$$

Hence, $(^{87}\text{Rb}/^{86}\text{Sr})_{\text{eff}} = 0.26$. This value characterizes the integrated mean $^{87}\text{Rb}/^{86}\text{Sr}$ ratio from the formation of the solar system $(^{87}\text{Rb}/^{86}\text{Sr} = 0.92)$ to the moment when it was eventually fixed in the Earth $(^{87}\text{Rb}/^{86}\text{Sr} = 0.09)$.

The relevant curve approximating the change of the $^{87}\text{Rb}/^{86}\text{Sr}$ ratio during the first $\sim 120\,\text{Ma}$ is shown in Fig. 7.8 (curve 1). It is plotted following the same rule as described in the previous (U–Pb) section.

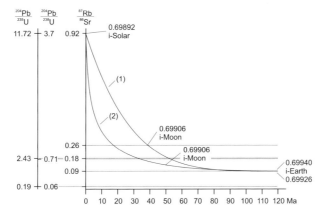

Figure 7.8. Evolution of $^{87}\text{Rb}/^{86}\text{Sr}$ ratio during first 120 Ma of the solar system history due to Rb hydrodynamic escape suggested by the evaporative accretion model, with initial $(^{87}\text{Sr}/^{86}\text{Sr})_{\text{i-Earth}}$ values: 0.69940 (curve 1) and 0.69926 (curve 2).

Accordingly, the Sr-isotope ratio changes from the solar initial (0.69892) to the accepted Earth initial (0.69940). Hence we can determine the time (t_M) when the Sr-isotope ratio, fixed in the Moon, could be achieved:

$$0.69906 = 0.69892 + 0.26 \cdot 1.07(e^{1.42 \cdot 10^{-11} \cdot t_M} - 1).$$

This equation gives $t_M = 37\,\text{Ma}$. Thus, on the basis of McCulloch's $(^{87}\text{Sr}/^{86}\text{Sr})_{\text{i-Earth}}$ value we find the fragmentation time to be about 40 Ma. In turn this means in our model that the lunar core began to form at least 10 million years after the fragmentation event, taking into account the 50 Ma constraint for the beginning of lunar core formation that follows from the Hf–W data interpretation.

Inversely, if the model suggesting the fragmentation occurred at 50 Ma is assumed then we will come to a different estimate of the $(^{87}\text{Sr}/^{86}\text{Sr})_{\text{i-Earth}}$ value:

$$0.69940 = 0.69892 + (^{87}\text{Rb}/^{86}\text{Sr})_{\text{eff}} 1.07 \cdot (e^{1.42 \cdot 10^{-11} \cdot 0.05 \cdot 10^{-9}} - 1).$$

In this case the average $(^{87}\text{Rb}/^{86}\text{Sr})_{\text{eff}} = 0.18$. The respective curve is also plotted in Fig. 7.8. Then:

$$(^{87}\text{Sr}/^{86}\text{Sr})_{\text{i-Earth}} = 0.69892 + 0.18 \cdot 1.07 \cdot (e^{1.42 \cdot 10 - 11 \cdot 0.12 \cdot 10^{-9}} - 1),$$

and $(^{87}\text{Sr}/^{86}\text{Sr})_{\text{i-Earth}} = 0.69926$. The last value is also of acceptable magnitude.

Thus both considered versions are plausible. It should be noted that the $^{87}\text{Rb}/^{86}\text{Sr}$ ratio fixed in the Moon (0.018) is not identical to that attained by the gas-particle cloud by the moment of its fragmentation. The latter determines the $(^{87}\text{Sr}/^{86}\text{Sr})_{\text{i-Moon}}$ value. The Rb could be quickly lost during the high temperature fragmentation process, but this could not be reflected in the $^{87}\text{Sr}/^{86}\text{Sr}$ ratio already fixed in the Moon. Additional correction might be done taking into account $\sim 30\%$ of additional material accreted by the Moon's embryo after fragmentation. Then the 50 Ma constraint is satisfied given $(^{87}\text{Sr}/^{86}\text{Sr})_{\text{i-Earth}} = 0.69916$. All values from 0.69916 to 0.69940 are plausible and within the limits of uncertainties of the present-day estimations of the Earth's initial Sr-isotope ratio. More precise determining of the fragmentation time in our model depends on more precise knowledge of the $(^{87}\text{Sr}/^{86}\text{Sr})_{\text{i-Earth}}$ value.

A most important aspect of the Rb-Sr systematic is that it is quite consistent with the concept of evaporative accretion and allows evaluation of one of the essential parameters of the process-time of fragmentation.

On the other hand it should be stressed that if the Earth and the Moon have different initial Sr-isotope ratios, which is most likely, this absolutely rules out the entire homogenization of the Earth's and the proto-lunar material suggested by the giant impact model.

In concluding this chapter we suggest a scheme of evolution of the gas-dust formation that illustrates the timing inferred from the above analysis of the isotopic systems (Fig. 7.9).

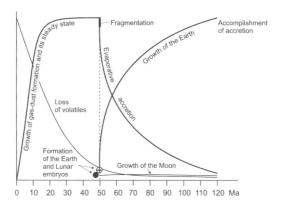

Figure 7.9. Qualitative scheme of evolution of gas-dust formation including phases of growth, steady state, fragmentation, and loss of mass during evaporative accretion. The latter is complementary to the growth of condensed bodies (Earth and Moon embryos), predominantly the Earth. The vertical scale is conditional.

We suggest that in the initial phase accumulation of the primitive material in particulated (gas-particle) form occurs. It is under contraction and loses volatiles but simultaneously can acquire new portions of the particulated material, so its mass is in more or less steady-state condition. In some period of time, contraction of this gas-dust cloud results in the collapse of its denser internal part. It happens in the period 40–60 million years after the beginning of growth of gas-dust accumulation (beginning of the solar system). The collapse is accomplished with fragmentation accompanied by the emergence of the Earth's and the Moon's embryos. During the period following fragmentation the mass of the initial vapor-particle cloud decreases as the mass of the Earth increases due to continued evaporative accretion. The process ends by about 120 Ma.

In the Part II we consider the dynamics of this evolution.

Part II

Dynamics

From the geochemical constraints and results presented in the previous chapters it follows that a realistic model of the origin of the Earth-Moon system should explain the formation of both Earth and Moon from a common source. If two bodies are derived from a common source then it means that at some point of the system history a kind of fragmentation took place. On the other hand the geochemical data shows that the current compositions of Earth and Moon are significantly different. Consequently the two bodies must have evolved differently after their separation. In other words the genesis of both bodies is the same but the subsequent evolutions are different. Dynamical analysis of the Earth-Moon system origin is the purpose of this part of the Book. Part II includes two chapters. The first, Chapter 8, considers the dynamic feasibility of fragmentation and emergence of proto-Earth and proto-Moon in the process of the rotational collapse of the gas-dust cloud. The second, Chapter 9, considers the subsequent process of accretion of the Earth's and the Moon's embryos, and explains the differences of their evolution after separation.

Chapter 8

Dynamical modeling of fragmentation of the gas-dust cloud

We start our dynamic consideration from the idea that the Earth-Moon system has been produced by fragmentation from a large but, in solar system terms, rather compact gas-dust cloud. The proto-Earth-Moon cloud formed as a collision and accumulation of smaller gas-dust clumps. The cloud was emplaced on a heliocentric orbit. The internal equilibrium of the cloud was maintained by the balance of the gravitational forces and the gas pressure forces produced by evaporation of volatiles from the dust particles. When the main sources of the vapor were exhausted the cloud began to collapse under its own gravitation. Due to conservation of the angular momentum the angular velocity of system rotation increased, which lead to the formation of separate condensed bodies. The model's peculiarity is that it suggests the formation of condensed bodies as a gas-dust cloud contracts by expelling volatile components. Therefore apart from the gravitational interaction, a gas dynamic component has to be taken into account.

This chapter describes the detailed analysis of the rotational collapse on the basis of the computer simulation using particles. The chapter is mainly devoted to dynamical and computational aspects of the problem, studying the influence of dynamical parameters on the cloud evolution. Material used in the construction of a computational model, the development of simulation algorithms, the implementation of numerical experiments, and the obtainment and processing of results will also be addressed.

The content of this chapter is based on the papers Galimov et al. (2005), Galimov and Krivtsov (2005), Le-Zaharov et al. (2005), Le-Zakharov and Krivtsov (2012), Mukhin and Volkovets (2004).

8.1 Computational modeling using particle dynamics

The particle dynamics method is a well-known computer simulation method (Hockney and Eastwood, 1988). From a computational viewpoint, it consists of trajectory calculations for a large number of interacting particles moving according to prescribed interaction laws. In the simplest case, where the particles are represented by mass points and the interaction is restricted to gravitational attraction, the problem of finding positions of all particles as functions of time is called the N-body problem. For $N = 2$ its analytical solution was obtained by Newton. For three or more bodies, no analytical solution exists in general terms; there are, however, methods for its approximate and computational solution. In the current research, since the formation of condensed bodies is of primary interest, the collisional dynamics of particles have to be considered, which significantly complicates the problem. The collisionless model including only the gravitational interaction allows only the formation of enhanced-density zones to be determined, but it cannot describe the formation of condensed bodies, such as the embryos of the Earth and the Moon. To describe the gas dynamic component of the system the gas-particle interaction forces and the forces of repulsion emerging during intense evaporation of material from the particle surface are assumed to be described by the corresponding terms introduced into the particle interaction law (Galimov et al., 2005). Thus it was necessary to take into account the forces of dissipation and repulsion of particles that emerge as they approach each other.

Thus computer simulation seems to be the only realistic tool for solving such a problem. However, to compute the trajectories of the particles, it is necessary to calculate the forces acting in the system. Their direct calculation leads to complexity $O(N^2)$ at each integration step. This is admissible for most problems of celestial mechanics where the number of bodies is usually not too large. However, when it comes to the particle method, this imposes severe constraints on the field of its application. However, there exist a number of alternative interaction force calculation algorithms that allow this problem to be solved. For the solution of the considered problem a modification of the Barnes-Hut algorithm (Barnes and Hut, 1986) was developed. The classical Barnes-Hut algorithm is based on the combination of particles into a tree structure by their geometrical arrangement to increase the speed of the gravitational force calculation. The distinction of the algorithm developed here (Le-Zakharov and Krivtsov, 2012) is that it efficiently works for both short-range and long-range forces and that it allows calculations to be performed for a significantly nonuniform distribution of particles without a decrease in performance.

The dynamics of the system is described by the particle dynamics method (Hockney and Eastwood, 1988; Krivtsov, 2007). The method consists of representing matter as an assembly of interacting particles, to which classical dynamics equations can be applied. In these equations, inertial terms are counterweighed by particle interaction forces, which are specified as known functions of particle positions and velocities.

The motion of the particles is described by the equations of Newtonian dynamics:

$$m\ddot{\mathbf{r}}_k = \sum_{n=1}^{N} \frac{1}{r_{kn}} f(r_{kn}, \dot{r}_{kn}) \underline{\mathbf{r}}_{kn}, \quad \mathbf{r}_{kn} = \mathbf{r}_k - \mathbf{r}_n, \quad r_{kn} = |\mathbf{r}_{kn}|; \quad k = 1, 2, \ldots, N;$$

(8.1)

where $\mathbf{r}_k$ is the radius-vector of the kth particle, m is the particle mass, N is the total number of particles, and $f(r, \dot{r})$ is the particle interaction force defined as:

$$f(r, \dot{r}) = \frac{A_1}{r^2} + \frac{A_2}{r^p} + \frac{A_3}{r^q}\dot{r}.$$

(8.2)

The first term in Equation (8.2) is the gravitational interaction force; therefore, the coefficient A_1 is defined as $A_1 = -\gamma m^2$, where γ is the gravitational constant. The second term is a short-range repulsion force arising at the moment of particle collision. Based on experimental data on the impact compression of a solid body (Altshuler et

Section 8.1 Computational modeling using particle dynamics

al., 1958; Glushak et al., 1992) the power coefficient of $p = 13$ can be taken. For computational reasons the smaller value of this parameter $p = 6$ will also be considered for 3D computations, since the equilibrium distance between similar particles at the moment of their collision and repulsion, i.e. at the moment $f(a, 0) = 0$, is equal to their diameter a, $A_2 = -A_1 a^{p-2}$.

The third term describes non-conservative interaction between particles. This interaction models energy loss at particle collisions and also viscosity, caused by particle interaction with the gas (vapor), and surrounding particles, which is an essential part of the model and will be discussed later. This term produces resistance for the relative motion of particles, including the gas drag effect and shear viscosity. Without this resistance no evolution in the system can be observed and the planet embryos would not be formed in the simulation.

Assuming that the second and the third term should change proportionally to each other with changes of the distance between particles, we get $q = p + 1 = 14$.

Using the proposed assumptions, we can transform Equation (8.2) to the following form:

$$f(r, \dot{r}) = \gamma \frac{m^2}{a^2} \left[\left(\frac{a}{r}\right)^{13} \left(1 - \beta \frac{\dot{r}}{r}\right) - \left(\frac{a}{r}\right)^2 \right], \qquad (8.3)$$

where the dissipation coefficient $\beta = -A_3/A_2 > 0$ is introduced. The dimensional coefficient β characterizes the energy losses during collisions. It is more convenient to use the dimensionless ratio β/β_* instead of this coefficient. Here, β_* is the value of β at which the oscillatory motion in the system of two particles interacting according to law (8.3) transforms into a nonoscillatory one:

$$\beta_* = 2\sqrt{11} \sqrt{\frac{a^3}{\gamma m}}.$$

It is assumed that the energy loss under the effect of the dissipative component of the interaction force is transformed into the particles' inner energy. The reverse influence of thermal effects on the system dynamics is not considered.

Initial conditions are defined by particle position and velocity. For 2D simulations we take the initial shape of the particle cloud to be that of a two-dimensional disk with the density of particle distribution varying as:

$$\xi(r) = \frac{3}{2} \bar{\xi} \sqrt{1 - (r/R_0)^2}, \qquad (8.4)$$

where $\xi(r)$ is the 2D density (number of the particles in a 2D unit volume), r is the distance to the center, R_0 is the disk radius, and $\bar{\xi}$ is the cloud average density. The distribution (8.4) allows a "solid-body rotation" (Fridman and Polyachenko, 1984) with the angular velocity defined as:

$$\omega_s = \sqrt{\frac{3\pi^2 \gamma \bar{\xi}}{4R_0}}. \qquad (8.5)$$

The model of "solid-body rotation" for the initial state of the cloud was chosen for the following reasons. Firstly, this is one of the simplest classical initial states for self-equilibrium rotating dust cloud, and is used by many authors (Hunter, 1963; Fridman and Polyachenko, 1984). Secondly, angular velocity lower than the velocity of equilibrium rotation allows for this initial state the collapsing solution, which preserves uniform rotation and the shape of the density distribution at all times during the collapse. This allows consideration of the simulation as a final stage of the cloud collapse from the much less dense state. Thirdly, the considered 2D distribution corresponds well with the 3D distribution considered in the next section. However, during preliminary simulations a lot of alternative initial conditions with various density and velocity distributions were considered. The main conclusion was that although the process of the cloud collapse can differ under different initial conditions, the final stage (formation of two bodies) depends weakly on the conditions selection.

The 2D model is selected for reasons of simplicity. It is generally accepted that the protoplanet disk is substantially flattened, therefore the 2D model can be considered as a relatively good approximation (Bryden, G. et al., 1999; D'Angelo et al., 2002; Canup and Ward, 2002; Youdin and Shu, 2002). Moreover, as will be shown below, the 3D computations demonstrate the similarity of the 2D and relatively flat 3D models.

At the initial moment, the cloud rotates around the center as a rigid whole. The angular velocity is ω_0. In addition, a random velocity vector simulating the chaotic character of the particles' motion is added to the velocity of each particle.

For a numerical computation, we have to specify the system parameters. Moreover, problem solution in a two-dimensional definition requires a transformation to a two-dimensional state of parameters that in reality describe three-dimensional objects. The similarity principle is observed by introducing a dimensionless parameter:

$$\alpha = \frac{K^2}{\gamma M^3 R_c}, \tag{8.6}$$

where K is the rotational momentum (moment of momentum),

$$R_c = \left(\frac{3M}{4\rho_c}\right)^{\frac{1}{3}} \tag{8.7}$$

is the radius of a solid body concentrating the total mass (M) of all the particles in the system, where ρ_c is the density of the condensed body.

The introduced dimensionless dynamic parameter α is proportional to the ratio of the rotational kinetic energy over the potential energy of the gravitational interaction within a system. In the case of the Earth-Moon system, the dimensional quantities and the corresponding dimensionless parameter α have the following values: $K = 3.45 \cdot 10^{34}$ kg $\cdot$ m$^2 \cdot$sec^{-1}; $M = 6.05 \cdot 10^{24}$ kg; $R_c = 6.41 \cdot 10^6$ m; $\gamma = 6.67 \cdot 10^{-11}$ kg$^{-1} \cdot$ m$^3 \cdot$ sec^{-2}; $\alpha = 0.0126$.

On the other hand the dynamical coefficient of similarity α (7) is proportional to the square of the initial angular velocity ω_0 (the parameter used as the initial condition for

Section 8.1 Computational modeling using particle dynamics

the computations):
$$\alpha = \frac{3\pi}{4}\left(\frac{R_i}{R_0}\right)^4 \frac{R_0}{R_c}\left(\frac{\omega_0}{\omega_s}\right)^2, \qquad (8.8)$$

where R_i is the inertia radius of the cloud. This formula allows obtaining similarity between the real and computational systems.

For 3D simulations the cloud particles get the initial coordinates uniformly randomly distributed inside the ellipsoid

$$\frac{x^2 + y^2}{R_0^2} + \frac{z^2}{h_0^2} = 1, \qquad (8.9)$$

where x, y, z are the axes of a Cartesian coordinate system, R_0 is the length of two semimajor axes, and h_0 is the length of the semiminor axis of the ellipsoid. The cloud rotates around the z axis. The geometrical sizes of the ellipsoid are calculated from the specified particle density in the cloud, the number of particles, and its axial ratio. The latter parameter is specified as the ratio $\varepsilon_{hR} = h_0/R_0$. The particle density in the cloud is specified via the mean distance between the nearest particles d_0. Thus, the initial distribution of the coordinates is defined by three parameters: d_0, N, and ε_{hR}. Parameters R_0 and h_0 can be expressed in terms of these from the formula for the ellipsoid volume:

$$V = \frac{4}{3}\pi R_0^2 h_0 = Nd_0^3, \qquad (8.10)$$

$$R_0 = \sqrt[3]{\frac{3}{4\pi}N\frac{1}{\varepsilon_{hR}}\left(\frac{d_0}{a}\right)}a, \quad h_0 = \varepsilon_{hR}\left(\frac{d_0}{a}\right)a. \qquad (8.11)$$

Projections of the particle coordinates to the plane x, y give the 2D distribution (8.4), which was used for the 2D case:

$$\xi(r) = \frac{3}{2}\bar{\xi}\sqrt{1 - (r/R_0)^2}, \qquad (8.12)$$

where $\xi(r)$ is the 2D density (number of the particles in a 2D unit volume), r is the distance to the center, R_0 is the disk radius, and $\bar{\xi}$ is the average 2D density. If the angular velocity of the cloud rotation is specified by the relation

$$\omega_s = \sqrt{\frac{3\pi}{4}\frac{\gamma M}{R_0^3}}, \quad M = Nm, \qquad (8.13)$$

where M is the total mass of the system, then for this distribution the gravitational forces in 2D are completely compensated for by the centrifugal forces, and the disk can rotate as a solid whole around its center (Fridman and Polyachenko, 1984) – i.e., execute solid-body rotation. In the three-dimensional case, there is no complete compensation (the z component of the gravitational force cannot be compensated by the centrifugal one), but ω_s can be used as a scale for the angular velocities of the ellipsoid rotation. Moreover for thin disks ($h_0 \ll R_0$) the considered 3D and 2D distributions are completely analogous.

The initial particle velocities are the sum of regular and random components. The regular particle velocity is $\omega_0 \times r$, which allows the initial rotation of the cloud as a whole with an angular velocity vector ω_0, where r is the particle radius vector. The random particle velocity components have a uniform random distribution inside an ellipsoid in velocity space that is specified by the value of its two equal semiaxes lying in the plane of rotation, v_{rand}, and the ratio of the third semiaxis to one of the two remaining semiaxes, ε_{vz}. Thus, the initial particle velocities are specified by the parameters ω_0, v_{rand}, ε_{vz}. The quantities d_0, ω_0, v_{rand} are dimensional and they are measured relative to the scale parameters a, ω_s, and $v_s = \omega_s R_0$, respectively. Here, a is the particle diameter (the equilibrium distance between the pair of interacting particles), ω_s is the angular velocity of solid-body rotation.

8.2 Computational technique

This section describes the necessary technique to perform the computations using particles for the problem of rotational collapse of the gas-dust cloud. Those who are mostly interested in the physics of the considered process can omit this section; however this information is essential for performing efficient computations in this area and allows better general understanding of the problem.

From the mathematical point of view, the problem is reduced to a solution of a Cauchy problem for a set of regular differential equations (8.1). However, when solving these equations directly, the number of necessary operations at each step of integration is proportional to N^2, which becomes a considerable obstacle for computation of large systems. The gravitational potential is a long-range one, excluding the use of a cut-off radius, which is successfully applied to solve similar problems in molecular physics. To solve equations (8.1) a modified Barnes-Hut algorithm (Barnes and Hut, 1986; Le-Zakharov and Krivtsov, 2012) is used. The algorithm is based on a hierarchical decomposition of the computation area into squares whose sizes increase in a geometrical progression with the distance to the particle under consideration. Use of this algorithm allows reducing of operation number to the value $N \log N$, significantly enlarging the size of the systems for which the computation can be feasible.

8.2.1 Classical Barnes-Hut algorithm

The hierarchical methods are most undemanding to various peculiarities of the physical model, in particular, to jumps in the distribution of particles. On currently available hardware, they allow computations to be performed for systems with more than 10^6 particles. Actually, there exist only two classical hierarchical algorithms – the fast

Section 8.2 Computational technique

multipole method and the Barnes-Hut algorithm. All of the rest are to some extent modifications and combinations of these with other force calculation methods.

Both algorithms are based on a Taylor expansion of the potentials for the groups of particles relative to the center of mass. Several terms of the Taylor series allow an approximate potential from the corresponding group to be quickly obtained. In turn, this makes it possible to construct a fast algorithm for calculating the forces. This approximation is usually called the multipole one in the literature. The only difference between the Barnes-Hut method and the fast multipole method is that the Barnes-Hut method is used to calculate the potential at the location of an individual particle and the fast multipole method is used to derive a compact expression to calculate the potential in some region of space.

The algorithm can be subdivided into the following steps:

1. Combining the particles into a tree data structure by taking into account their proximity to one another. There exist implementations for building a tree by combining the groups of particles (the nearest particles are combined into pairs to form nodes, the pairs are then also combined between themselves, and so on). However, this is usually done just through hierarchical partitioning of the space into cubic cells. For the two-dimensional case, an example of such partitioning is shown in Fig. 8.1. The cells correspond to the tree nodes; the particles in them correspond to the leaves.

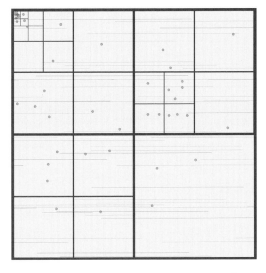

Figure 8.1. An example of a hierarchical partitioning of the space into cells for the 2D case.

2. To calculate the net force acting on an arbitrary particle, the tree is traversed from the root. When the next node is reached, the subsequent calculation is performed according to the following scheme:

(a) if the node is terminal, then the force exerted from this node is just added to the result;

(b) if the node is not terminal, then an approximation can be calculated for the potential produced by the particles of this node. It is checked whether this approximation will be accurate enough.

– if yes, then the approximation is calculated and the traversal of this tree branch is finished;
– if not, then step 2 is repeated recursively for all child nodes.

Let us consider step 2b of the algorithm in more detail. The potential produced by a group of particles (node descendants) can be expanded as a power series in $1/r$ with coefficients dependent on θ and φ, where r, θ, and φ are the spherical coordinates relative to the node's center of mass. To derive an approximate expression for the potential, it will suffice to take several terms of the series, usually from one to eight. The most popular methods are those limited to the zero approximation order, i.e., including only the first expansion term. In this case, the field of the point potential from the node's center of mass is obtained.

Before the force acting on a particle is calculated using this approximation, it is checked whether this approximation will be accurate enough. The algorithm is based on the fact that the multipole expansion coefficients for each node of the tree can be calculated from the coefficients of its child nodes. This makes it possible to recalculate them in one pass over the tree from bottom to top in a linear time.

The decision criterion in step 2b is usually called the Multipole Acceptance Criterion (MAC) in the literature. The MAC is often described by a quantity θ – the so-called opening angle. Physically, θ is the maximum angle at which the cell must be seen from the location of the particle for which the force is calculated for the multipole approximation to be used. The following three types of MAC are most popular:

1. **Barnes–Hut (BH) MAC:** $s/r < \theta$, where r is the distance from the particle to the center of mass of the cell, s is the cell size.

2. **Min-distance (MD) MAC:** $s/r < \theta$, where r is the distance from the particle to the cell boundary, s is the cell size.

3. **Bmax MAC:** $b_{max}/r < \theta$, where b_{max} is the maximum distance from the center of mass of the cell to its boundary, r is the distance from the particle to the center of mass of the cell.

If the MAC condition is met, then the multipole approximation is considered admissible in this case. All three listed criteria are analyzed in detail by Salmon and Warren (1994). The authors showed that the MD MAC has an advantage over the remaining criteria in the maximum possible relative error in calculating the acceleration of a specific particle.

Salmon and Warren were the first to try to rigorously calculate the a priori error introduced by the multipole expansion when using different criteria. It turned out that the relative error in calculating an approximate force for the purely empirical Barnes-Hut criterion could reach unity if $\theta > 1/\sqrt{3} \approx 0.577$. In this case, $0.7 < \theta < 1$ is commonly used. In the two-dimensional case, $1/\sqrt{3}$ in the first inequality is replaced by $1/\sqrt{2} \approx 0.707$. In addition, Salmon and Warren were able to provide an example quite possible in practice, where an almost 100 % relative error in calculating the forces by the BH method is realized at bad θ.

Before Salmon and Warren, Hernquist (1987) obtained a posteriori estimates of the error for the Barnes-Hut method by comparing the results of calculations with those obtained through an exact calculation of the forces. He obtained Gaussian distributions of the error over the particles. The root-mean-square relative error in this case did not exceed 1–2 % (depending on the calculated parameters).

8.2.2 Data structure

The main criteria determining the overall performance of the force calculation module when using the Barnes-Hut algorithm are:

1. building speed;

2. memory size;

3. traversal speed;

4. adaptation to a nonuniform particle distribution.

Equilibrium should be kept between these four properties. A strong skewing in favor of a particular property will most likely have a negative effect on the overall performance.

Here an implementation of the algorithm based on a hierarchical partitioning of space is described. A particle and a cell are used as two key concepts. A particle is represented by its coordinates and velocity. A cell is determined by the corresponding level in the partitioning hierarchy and three integer coordinates (the order numbers along the three axes of space). Some additional data needed for the numerical scheme to work (e.g., the coordinates of the center of mass and the multipole expansion coefficients) correspond to both cell and particle. Depending on the implementation, a cell can also store a set of its particles and references in its child cells.

A fixed number of hierarchy levels are used in the implementation of the algorithm under consideration. This means that there exists some maximum – *base* – hierarchy level and the fragmentation occurs up to this base level in the regions of space where at least one particle is present. This makes it possible to distribute the particles over the base level in one pass over them at the tree building step and then to build the rest of the tree over this level.

The tree traversal can be accelerated significantly if the lists of particles in the cells are stored not only at the base level but also in the entire tree. However, if this is done directly, then the lists will duplicate each other. After all, it is obvious that the parent cell contains the set of particles located inside its child cells. Therefore the unified list of particles is sorted out in such a way that for any cell of any level all of the particles belonging to it lie in this list sequentially one after another.

Since the particles are initially distributed over the base level, the cells should also be properly sorted out. If this is done not only at the base level but also at all the remaining ones, then the following will be obtained. For an arbitrary cell, its eight child cells will be in memory together, i.e., sequentially one after another. Their child cells will be stored in exactly the same way and so on. The corresponding cell numbering for the two-dimensional case is demonstrated in Fig. 8.2.

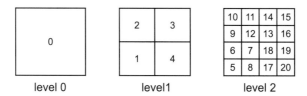

Figure 8.2. An example of the cell numbering for the 2D case.

At first glance, the procedure for calculating the number of a cell from its integer coordinates and vice versa is too laborious for this numbering scheme to be used in practice. However, it turns out to be just a set of bit operations. Fig. 8.3 shows the way of calculating the number $n_l(c)$ within level l of cell c with coordinates i, j, k. Thus, the bit representation of $n_l(c)$ is formed if we sequentially take one bit from each cell coordinate. Subsequently, the global cell number is calculated from the formula

$$n(c) = \sum_{s<l} m_s + n_l(c), \qquad (8.14)$$

where m_s is the number of cells at level s.

Let us describe the tree building process. At the initialization step, the number of hierarchy levels is determined and auxiliary data structures are built, but the tree itself is rebuilt during each force calculation. It is built from bottom to top – from the root to the leaves. The particles are distributed over the cells of the lowest (*base*) hierarchy level. Subsequently, the set of particles is sorted as described above. Next, the upper part of the tree is built, with the empty cells being ignored. The set of particles belonging to each cell is associated with it, which is basically only part of the general list.

The original method of data storage in memory has a number of undeniable advantages. Let us briefly list them once again:

– an easy-to-implement method of particle distribution over the tree;

– fast access to the particles of any cell: for each cell, it will suffice to store the number of its first particle and their number;

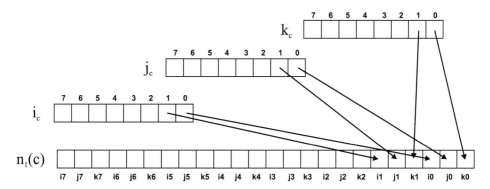

Figure 8.3. Calculation of the cell number for the 2D case

- fast access to the descendants of any cell: the descendants of one level for any cell are located in memory sequentially, which optimizes the operation of cache memory and speeds up some operations in parallelization;

- memory saving: since the child cells for any cell are stored in a group of eight pieces, it will suffice to store the reference only to the first descendant in the node;

- fast traversal of the entire tree from top to bottom or from bottom to top by a simple pass over the set of cells.

8.2.3 Tree traversal

In the classical Barnes-Hut algorithm, the force acting on each particle is calculated separately. In this case, the tree is traversed starting from the root. The Multipole Acceptance Criterion (MAC) is applied in each node. For two particles that are close to each other, the order of tree traversal is virtually the same. In the fast multipole method, the potential from a group of particles is approximated at once for some region of space. However, this requires the additional operation of "distributing" the potential from cells to particles after the tree traversal.

In the protoplanetary cloud model of the Earth-Moon system described above, stable clusters of particles are formed at some step of computation. These clusters contain up to 80 % of the system's total mass and the density in them exceeds the mean density of the configuration by one order of magnitude. It emerged in the course of our numerical experiments that performing a computation for such systems using the classical Barnes-Hut algorithm is inefficient. The cluster particles are too close to one another. For this reason, when the forces for such particles are calculated, the tree traversal is made fairly "deeply" and takes considerable time. The part of the tree where the cluster is located is examined down to the lowest level and this is repeated as many times as there are particles in the cluster.

The fast multipole method is to some extent free of this problem, but in practice this algorithm is efficient for a larger number of particles than can be processed by today's computers (Capuzzo-Dolcetta and Miocchi, 1997). In addition, because of the presence of a short-range component in the interaction law and because of the essentially nonuniform distribution of particles, the physical model used limits the number of particles in the simulated configurations still further.

The ideas of the two algorithms can somehow be combined if the acceptance criterion is applied simultaneously for groups of particles. The particles lying near each other will be "dragged" through the tree of cells together, separating only at the lowest hierarchy levels. This will reduce the number of MAC application and cell opening operations.

Suppose that we want to calculate not the force acting on each individual particle but all forces in the system at once. Since the computational domain is represented by one cell – the tree root – it can be imagined that we want to calculate the influence of this cell on itself. Let us transform the acceptance criterion. At the input, it will receive two cells, the *requesting* and *requested* ones. The MAC result says whether the requested cell could be opened when calculating the forces for the particles located in the requesting cell using the classical Barnes-Hut algorithm. Using this algorithm, we can traverse the tree in such a way that a pair of cells rather than a particle and a cell, as in the Barnes-Hut method, will always be analyzed:

1. The tree traversal is started from the "root-root" pair.

2. The acceptance criterion is applied in each node:

    (a) if the multipole approximation is possible, then an approximation of the force from the particles of the right cell is calculated for each particle of the left cell;

    (b) if not, then it is checked whether both cells are the base ones:
     – if both are the base ones, then the interaction forces between the particles in the cells are calculated directly;
     – otherwise, the larger cell is opened and step 2 is repeated recursively for all the pairs of cells obtained as a result of the opening.

To rule out the opening of cells with a small number of particles, a special *balancing constant* – the minimum number of particles at which the cell can be opened – is added to the parameters of the algorithm. In step 2b of the algorithm, apart from checking the cells for being terminal, checking for the number of particles is also added.

8.2.4 Parallelization scheme

There exist two ideas of parallelization: 1. to distribute the particles between the processes and 2. to distribute the computational domain between the processes. Here, the second idea was taken as a basis. The point is that the forces between close-lying particles are calculated directly. For particles located slightly farther apart, a fairly accurate approximation is used (small cells are taken). The size of the cells used for the approximation increases with distance between the particles, i.e., detailed information near the particle and less detailed information far from the particle is required to calculate the force acting on it. Consequently, if all particles of each process are near one another, then it will be necessary to send much less information between the processes (Fig. 8.4).

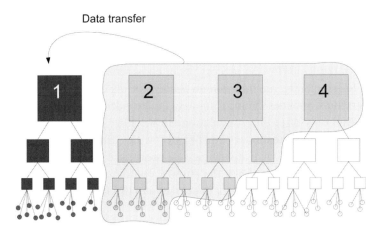

Figure 8.4. Data transfer during parallel computing.

Let us consider in more detail how the processes interact during parallel computing. At the initialization stage, the zero-rank process distributes the computational domain by enumerating all cells of the subdomain level in order of decreasing number of particles. In this case, the next cell is given up to the least loaded process.

For a small number of particles per processor, a situation may well arise where one of the processes will get part of the computational domain with an excessively large number of particles. The calculation of the forces in it will be slower and the remaining processes will wait for it; in this case, the overall efficiency of the computations will decrease. In the existing implementation of the algorithm, this problem is solved by increasing the subdomain level. This turned out to be sufficient at the stage under consideration. However, the development of some sort of adaptive partitioning method is planned in future. For reference, it can be added that when a configuration made of one million particles was computed on 64 processors, a level containing $2^{18} \approx 260\,000$ cells was taken as the subdomain one.

The subdomain cells obtained by the processor during the distribution can be combined into larger ones. After this procedure, the computational domain for each process is represented by a list of non-overlapping cells of various levels; we will call them *quasi-subdomain* cells. The force calculation algorithm undergoes virtually no changes. The only difference is that the tree traversal is started not from the "root-root" pair but from the list of "quasi-subdomain cell-root" pairs.

Information about the cells belonging to other processes may be needed in the force calculation. To understand what information is necessary, all quasi-subdomain cells are enumerated. The lists of requests to other processes can be determined by applying the MAC to the quasi-subdomain cells as the requesting ones and to all non-local cells as the requested ones. Lest such a complete enumeration be made, it is worth traversing the non-local cells over the levels from top to bottom, discarding the entire subtree together with the unnecessary cells.

One integration step occurs as follows. Once the tree has been built, the multipole coefficients in the local (for the process under consideration) part of the tree are recalculated. Subsequently, the information needed for the coefficients in the common part (the part that combines the local subtrees) to be calculated in each process is recalculated. And only after their calculation are the forces calculated (separately for the particles in the computational domain and for the escaped particles). The choice of the time step used to recalculate the particle coordinates and velocities depends on the magnitude of the forces. When the particle coordinates have been updated, the particles are redistributed between the processes and the step is finished.

8.2.5 Comparative analysis of the performance

It is rather difficult to objectively compare the performance of various implementations of the Barnes-Hut method. The published data on speed measurement are often incomplete. The speed of computations depends largely on the computation parameters affecting the accuracy of the calculations. In addition, the operations were performed on various hardware platforms and, occasionally, it is just impossible to find complete data on their performance. The "nonuniformity" of the particle distribution in space also plays a very important role. As a rule, they are distributed uniformly or according to some weakly changing law at the initial instant of time. In contrast, the final configuration can be determined by a number of factors and it is usually impossible to predict it before the start of calculation.

Table 8.1 presents the results of our tests (Le-Zakharov and Krivtsov, 2009) for some known implementations of the Barnes-Hut algorithm. The results were obtained on the MVS-100K supercomputer of Joint Supercomputer Center of the Russian Academy of Sciences (peak performance of 140 Teraflops). According to the table, faster results than Le-Zakharov and Krivtsov (2009) were obtained only by Makino (2004) and Hamada et al. (2009). Makino (2004) obtained very successful results by the construction of special-purpose computers for the calculation of gravitational forces (GRAPE). The results of Hamada et al. (2009) were obtained using graphical processors (GPU) that allowed a large number of similar operations to be optimized with a higher efficiency. However, in Makino (2004) and Hamada et al. (2009) only the gravitational forces were present in the model. The absence of short-range forces

Table 8.1. Performance data for some of the best known implementations of the Barnes-Hut algorithm in comparison with the current project. is the number of particles, is the time spent on one integration step, is the number of processors, is the ratio of the total processor time per step to the number of particles in the system.

Author/project	Year	Initial confoguration	N	t_s, s	P	t_p, μs
Le-Zakharov and Krivtsov (2009)	2009	ellipsoid (model of rotating gas-dust cloud)	1M	70	1	70
Le-Zakharov and Krivtsov (2009)	2009	ellipsoid (model of rotating gas-dust cloud)	1M	1.37	64	88
Le-Zakharov and Krivtsov (2009)	2009	ellipsoid (model of rotating gas-dust cloud)	1M	0.98	128	125
Le-Zakharov and Krivtsov (2009)	2009	ellipsoid (model of rotating gas-dust cloud)	10M	6.7	256	172
Le-Zakharov and Krivtsov (2009)	2009	model of gas-dust cloud, final stage	1M	6.9	64	442
Makino (2004)	2004		2M	3	12+1	20
Anh and Lee (2008)	2008	Plammer distribution	2.56M	20	32	250
Stock et al. (2008), CPU	2008	Cube	500 K	251.5	2	1006
Stock et al. (2008), CPU+GPU	2008	Cube	500 K	14.9	3	89
Li et al. (2008)	2008	8 spheres	1M	345	1	345
Belleman et al. (2007)	2007		1M	733	1	733
Hamada et al. (2009) CPU	2009	Sphere	1M	635	1	635
Hamada et al. (2009) GPU	2009	sphere	1M	7.94	1	8
Hamada et al. (2009) GPU	2009	sphere	1M	0.13	128	17

has a positive effect on the speed of computations. It is also important that the most interesting and complex task is to achieve an acceptable performance for the clustered (or highly nonuniform) configuration forming at the final stage of the computation. For the current computations, the computation time of one step for a configuration made of a million particles on 64 processors changed from 1.37 to 6.9 seconds. Makino (2004) and Hamada et al. (2009) did not include such data and therefore no comparison can be made.

In Fig. 8.5, the speed of computations is plotted against the number of particles (Le-Zakharov and Krivtsov, 2009). The studies are carried out in the range of up to 10^6 particles, because it was decided at the design stage to restrict the project to this number. Our tests showed that a predominantly linear dependence is observed in the investigated range. This corresponds to the asymptotic estimate $O(N \log N)$ for the Barnes-Hut algorithm; $\log N$ has no significant effect for such a small number of particles.

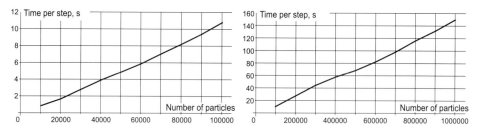

Figure 8.5. Time spent on one integration step. The number of particles changes in the range from to (left plot) and from to (right plot). The computations were performed on an ACER Aspire 5030 computer, 512 Mb RAM, AMD Turion 64 Mobile Technology ML-30, 1.59 GHz under MS Windows XP.

8.2.6 Results of the computations, 2D-model

Figure 8.3 shows the results of computation of the rotational collapse for the dust cloud in oblique projection. The number of particles is $N = 10^4$, the initial radius of the cloud is $R_0 = 5.51 R_c$, the initial angular velocity of the cloud rotation is $\omega_0 = 0.70 \omega_s$, where ω_s is the angular velocity of the solid-state rotation (Equation (8.13) above). The random velocity component in the initial configuration is taken from a uniform random distribution with the maximum value $0.68 \omega_s R_0$. The time corresponding to the successive frames in Fig. 8.3 is measured in the units of $T_s = 2\pi/\omega_s$, which is the period of solid rotation for the initial configuration. The spectral color scale from blue to red illustrates the temperature change from the lowest to the highest temperature in the system.

One can see that the collapse is accompanied by a formation of two density conglomerations of different sizes, which are being transformed into condensed bodies. These conglomerations feature a relatively elevated temperature.

The number of the density conglomerations depends on the initial angular velocity of the cloud. However, the considered system is highly stochastic due to the random distribution of the particle co-ordinates and velocities in the initial configuration. Therefore for the same values of the initial angular velocity different scenarios can be realized – for example in Fig. 8.6 the collapsed configurations corresponding to $\omega_0 = 0.70 \omega_s$, $T = 0.84 T_s$ and different random number generations are shown.

To analyze the system behavior depending on the initial angular velocity a great number of computational experiments accompanied by statistical data processing is required. In Fig. 8.7 the results of more then 600 numerical experiments of the rotational collapse of the dust cloud are summarized (Le-Zaharov et al., 2005; Le-Zakharov and Krivtsov, 2012; Mukhin and Volkovets, 2004. The computations were performed on the multiprocessor computational system RSC4, M. V. Keldysh Institute for Applied Mathematics, Russian Academy of Sciences). The different values of the relative initial angular velocity vary within the limits from $0.28 \omega_s$ to $0.85 \omega_s$ with the

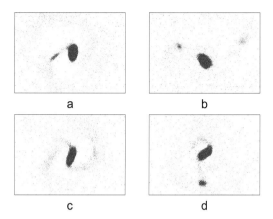

Figure 8.6. Results of the rotational collapse for $\omega_0 = 0.70\,\omega_s$, $T = 0.84\,T_s$; a)–d): configurations corresponding to different random number generations.

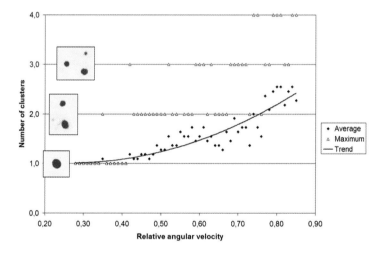

Figure 8.7. Dependence of the number of clusters on the initial angular velocity.

step $0.01\,\omega_s$. For each value of the angular velocity the series of 11 independent computations based on different random number generations were performed. The graph shows the maximum number of developed clusters for each value of the angular velocity (triangles) and the average number of clusters in each series (diamonds). The clusters are determined as particle ensembles with a distance to the nearest neighbor of less then $2a$ at $T = 0.84\,T_s$.

From the graph it can be seen that no fragmentation takes place if the relative angular velocity ω_0/ω_s is lower then the critical value of approximately 0.42. An increase of the angular velocity leads to an increase of the probability for the two bodies' development: for the value $\omega_0/\omega_s = 0.6$ the development of three bodies becomes quite probable, for $\omega_0/\omega_s = 0.8$ four bodies can be formed. The graph shown in Fig. 8.8 illustrates the dependence of the mass ratio between the second-sized and the first-sized clusters. According to the graph if the angular velocity is close to the critical one then the size of the second cluster is very small. Increase of the angular velocity leads to an increase of the second cluster size, so that for the velocity $\omega_0/\omega_s = 0.8$ almost similar-sized clusters can be formed. Fig. 8.9 shows typical results of the rotational collapse for the different values of the initial angular velocity.

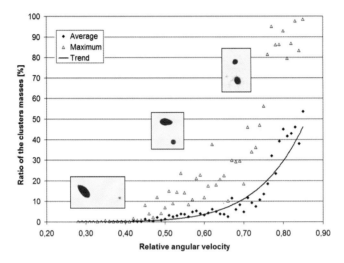

Figure 8.8. Dependence of the mass ratio between the second-sized and first-sized clusters on the relative angular velocity.

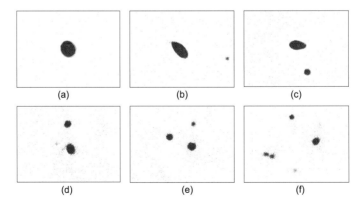

Figure 8.9. Results of computation of rotational collapse for different values of the initial relative angular velocity ω_0/ω_s: a) 0.29, b) 0.42, c) 0.54, d) 0.76, e) 0.80, f) 0.85.

Section 8.3   Evaporation of the particles as an important factor of fragmentation   111

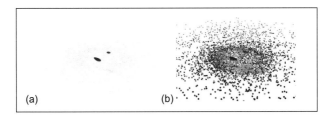

Figure 8.10. The cloud of particles in space surrounding the condensed bodies immediately after their formation ($t = 1.07 \, T_s$): a) original view; b) particles radii are increased for better visibility.

An important feature of a collapse with a chaotic component of particle velocity is that a significant fraction of particles stays scattered over the space after the condensed bodies are formed (Fig. 8.10), while the temperature of these scattered particles is substantially lower than that of the condensed bodies. Later on, the slow process of accumulation of scattered particles by the condensed bodies is realized, which takes many orders of magnitude more time that the process of the condensed bodies' formation.

8.3   Evaporation of the particles as an important factor of fragmentation

The evolution scenario illustrated in Fig. 8.3 could have made a basis for representation of the formation of the Earth-Moon double system but for one very important and highly disagreeable circumstance. The problem is that the angular momentum of the actual Earth-Moon system is insufficient to develop a rotational instability. Kinetic momentum of the actual Earth-Moon system corresponds to the value $\omega_o/\omega_s = 0.08$ (as one can infer from Equation 8.13); this is much smaller than the critical value $\omega_o/\omega_s = 0.42$ required for rotational instability (Fig. 8.7). Hence, we seem to have obtained yet another unsuccessful scenario of the Moon formation as a result of the rotational instability of the initial system.

However, the situation changes if the evaporation process is taken into account.

The evaporation process generates an effect of one more force that should be accounted for in the equation of motion (8.1). Note that it is not the value of the evaporation intensity that matters, but the increase of the evaporation intensity under the conditions of an equilibrium state of the vapor/condensed phase system. With such an equilibrium (reversible) increase of the intensity, the isotope fractionation is characterized by the thermodynamic isotopic effect. The increase of vapor flow from the surface of the particle, caused by pressure from saturated vapor, produces a repulsive force that can be approximately defined as:

$$f_V = \frac{\pi \nu \upsilon a^4}{16 r^2}, \tag{8.15}$$

where ν is the additional mass of the matter evaporated from a unit of the particle surface per time unit, and υ is the average velocity of molecules torn off from the particle surface. Hence the repulsion force due to the evaporation flow is inversely proportional to the squared distance between neighboring particles. Therefore, in local approximation it can be summed with the gravitational interaction force:

$$f = f_\gamma - f_\nu \simeq \left(\gamma - \frac{9\nu\upsilon}{4\pi a^2 \rho^2}\right)\frac{m^2}{r^2} \simeq \gamma' \frac{m^2}{r^2} \tag{8.16}$$

Let us note that the dissipative influence of the vapor surrounding the particles and leading to the gas drag effect is already taken into account by the third term of the equation (8.2).

As mentioned above, the appearance of rotational instability is determined by the value of the dimensionless dynamic parameter α, which includes γ. Having taken the effective value γ', we can determine the evaporation intensity level sufficient to produce a rotationally instable system that has the parameters of the Earth-Moon system.

A rotational instability resulting in formation of two separate bodies appears when the value of the angular velocity ratio is approximately between $\omega_0/\omega_s = 0.42$ and $\omega_0/\omega_s = 0.76$, i.e. 5.3–8.5 times higher than the values of ω_0/ω_s calculated for the Earth-Moon system. It means that the value of γ' is to be at least 28 times smaller than the value of γ.

It follows from Equation (8.10) and $\upsilon = \sqrt{8RT/\pi}$ (Kuchling, 1980), where R is the specific gas constant and T is the absolute temperature, that the critical value of $\nu = \nu_1$ required for the appearance of rotational instability is

$$\nu_1 = C_1 \frac{\gamma \rho^2 a^2}{\sqrt{RT}}, \quad \text{where} \quad C_1 = \frac{4}{9}\sqrt{\frac{\pi^3}{8}} \approx 0.87. \tag{8.17}$$

From the above formula it follows that the ν_1 value decreases with temperature growth and particle size decrease. The value $\nu = \nu_2$ required for rotational instability can be determined by the formula

$$\nu_2 = C_2 \frac{\gamma \rho^2 a^2}{\sqrt{RT}}, \quad \text{where} \quad C_2 = \left(1 - \frac{\gamma'}{\gamma}\right)C_1 \approx 0.84.$$

For the particles with the size of meteorite chondrules ($a \simeq 1$ mm), temperature of the order of 10^3 K and density between 10^3 to $2 \cdot 10^3$ kg/m^3 the ν_2-value has an order of 10^{-13} kg/m$^2 \cdot$ sec (for $R \simeq 400$ J/kg/K). Hence, a small increase in evaporation intensity can result in fragmentation of the collapsing cloud of particles. The rate of decrease of particle mass is calculated from the following equation:

$$\frac{dm}{dt} \simeq -\pi C_2 \frac{\gamma \rho^2 a^4}{\sqrt{RT}}. \tag{8.18}$$

The time required to evaporate a part of the particle mass Δm under an evaporation rate given by Equation (8.17) can then be calculated by the formula

$$t = \frac{1}{2C_2} \frac{\sqrt{RT}}{\gamma \rho a} \left[\left(1 - \frac{\Delta m}{m}\right)^{-1/3} - 1 \right]. \tag{8.19}$$

For instance, a 40 % decrease of the mass of a particle under conditions mentioned above would take $3 \cdot 10^4$ to $7 \cdot 10^4$ years.

8.4   3D-model of evaporative fragmentation

In the previous section the rotational collapse of the gas-dust cloud resulting in formation of two condensed bodies was analyzed in 2D. Possibilities of 2D simulation of such 3D processes are based on the ideas that due to the rise in the angular velocity in the collapse process the cloud should be considerably flattened and in this case a 2D approach gives reasonable approximation of the real 3D process. The similarity principles were used to obtain the correspondence between the 2D model and the 3D object under investigation. However, the final validation of the obtained results cannot be done without real 3D modeling. Therefore in this section the 3D model of the process is considered and the corresponding simulation results will be analyzed. It will be shown that in general all the main conclusions obtained for 2D are confirmed by 3D simulations. However, many additional aspects should be taken into account for the proper construction and evaluation of a 3D model.

8.4.1   Simulation parameters and numerical experiments

The initial conditions for the 3D simulations can be reduced to selecting numerical values of 6 parameters: d_0, N, ε_{hR}, ω_0, v_{rand}, ε_{vz}, namely the mean distance between the nearest particles, the total number of particles, the axial ratio for the cloud, the initial angular velocity, the maximum random velocity, and the axial ratio for the random velocity ellipsoid. Strictly speaking, the values of these parameters are indeterminate and one of the problems is to investigate the behavior of the system as a function of these values.

From physical considerations, it is desirable that the velocities and coordinates of the system at the start of simulation should correspond to one of its evolutionary stages. This means that at least there should be no qualitative jump in the pattern of cloud motion immediately after the start of simulation. In other words, the stage of relaxation of the system to the characteristic motion to be investigated should not last long, lest distortions be introduced into the process being simulated. As applied to the problem in question, this means that the separation of a significant fraction of particles or, for example, prolonged, gradually decaying oscillations of the cloud "thickness" (its sizes along the axial symmetry axis) suggests a poor choice of initial conditions. In a number of related molecular dynamics problems, this problem is solved as follows. Once the initial values of the dynamical variables have been specified, some time is allocated for the system to come to equilibrium and only afterward is allowance for the effects being investigated "switched on" in the model and the quantities of interest measured. The establishment of equilibrium is recorded using an appropriate physical criterion, for example, by the establishment of a Maxwellian velocity distribution or by equalization of the particle number densities in various volume elements of the system. Thus, during a numerical experiment, the simulation stage is preceded by the stage of preparation of the initial conditions. For the problem under consideration, this approach does not seem to be easily realizable, because the initial relaxation cannot be separated from the clustering process that is of interest. Thus, the only criterion for suitability of the chosen initial conditions is the absence of "undesirable" motion of the system at the initial instant of time.

In the two-dimensional case where the cloud is flat, the motion of the system "as a whole" at the initial instant of time depends on ω_0 and v_{rand} at fixed d_0 and N. At high values of the initial angular velocity ω_0 the particles rapidly fly away from the center of rotation over the entire space. At its low values, the particles "fall" to the center of mass of the cloud. As a result, a considerable number of particles collide, which again causes a certain fraction of particles to fly apart. The parameters of the initial conditions should be chosen so as to avoid the appearance of these effects.

The stability of two-dimensional collisionless gravitating systems has been extensively studied in the literature (Fridman and Polyachenko, 1984). The distribution law and the angular velocity (see Equations (8.12) and (8.13) above) at which the gravitational and centrifugal forces compensate each other and the disk executes the so-called solid-body rotation, being in a state of stable equilibrium, have been found for an infinitely thin disk. In the 3D case, the situation is much more complex. Equilibrium distribution laws have been found only for spherically symmetric configurations (Fridman and Polyachenko, 1984). The problem is that the centrifugal forces cannot compensate the vertical (along the axial symmetry axis) component of the gravitational forces. Therefore the vertical dimension of the cloud can be supported only by the vertical component of the random particle velocities. Thus, to specify an equilibrium cloud configuration, appropriate relations between ε_{hR}, ω_0, ε_{vz} at fixed d_0, N should be chosen properly.

The numerical experiments showed that the cloud collapsed along the axial symmetry axis at fairly large ε_{hR} (the ratio of the cloud thickness to its diameter) or small v_z. This results in strong oscillations of the cloud thickness along the vertical direction (Fig. 8.11a). On the other hand, a strong increase in the random component of the initial velocity v_{rand} leads to a rapid separation of particles in the radial direction. Another artificial form of cloud oscillations can be observed if initial vertical velocities are chosen between two maximum values directed up and down – in this case the cloud temporarily separates in two layers (Fig. 8.11b). The best result was obtained for the initial parameters

$$\varepsilon_{hR} \leq 0,15, \quad \varepsilon_{vz} = 1,0, \tag{8.20}$$

Section 8.4 3D-model of evaporative fragmentation

i.e., where the initial configuration is nearly flat and the maximum axial random velocity component is equal to the radial one (an isotropic velocity distribution, the random velocities are uniformly distributed in a sphere of radius v_{rand}). In this case, the oscillations disappear and the cloud begins to regain its shape (Fig. 8.11c).

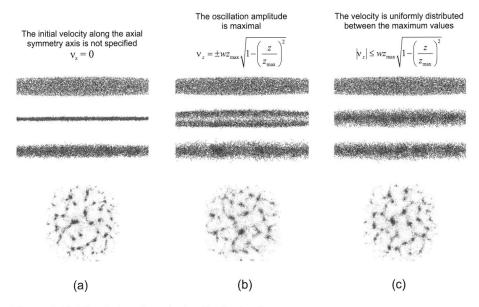

(a) (b) (c)

Figure 8.11. The choice of a velocity distribution law.

The initial cloud density (being set by the parameter d_0) is taken to be approximately one order of magnitude lower than the density of the condensed body composed of the same particles. The values of the parameters ω_0 and v_{rand} are considered as the main factors affecting the results of calculations. From common-sense considerations, it can be assumed that the initial angular velocity ω_0 must be in the range from $0.5\omega_s$ to ω_s, while the random velocity component must not exceed the range from $0.1\,v_s$ to v_s.

Thus, at this stage, the problem is reduced to investigating the dependence of the cloud evolution on four independent parameters – ω_0, v_{rand}, β, and N, namely the angular velocity of the cloud, the maximum random velocity, the dissipation coefficient, and the total number of particles. The desired results of the simulation are the mean number of formed clusters, the ratios of their sizes, the mean condensation time, the mean temperature of the dust and formed bodies, and some others. Let us note that the main interest focuses on statistical results, in other words the result of simulating any individual initial configuration is of little importance, because it depends strongly on the choice of the random distribution of the particle coordinates and velocities. In addition, the approximate force calculation algorithm and numerical integration introduce some error, which can strongly affect the result of calculating an individual configuration. Statistical data, on the other hand, is less at risk in the process of calculation.

The mean number of formed bodies is the most important result of the calculations. For whatever future studies are undertaken, it will first be necessary to ascertain the dependence of this quantity on the set of initial parameters and to identify the domain of ω_0, v_{rand}, β, and N in which the formation of two planetesimals will be most probable. Intuitively, it can be assumed that the angular velocity ω_0 and the random velocity v_{rand} will have a direct influence on this result. However, the influence of the two remaining parameters is not ruled out either.

It is worth discussing separately the choice of the parameter N. The values of this parameter are limited by the physical adequacy of the model from below and by the power of modern computers from above. Clearly, it would be ideal to consider the model closest to the real object being investigated which is the protoplanetary gas-dust cloud. Accordingly, the number of dust particles involved in the process is computationally huge. It is obviously impossible to perform such calculations using the real number of particles. We can only hope for a similarity in the system's behavior when the number of the particles is changed. The essence of this approach consists in the following. Suppose that we obtained the results for thousands, tens and hundreds of thousands, and millions of particles. If these results coincided, then it can be accepted that they will also be correct for a larger number of particles.

8.4.2 Modification of the parameters (interaction potential, angular and random velocities, and dissipation coefficients)

Preliminary calculations show that the bodies being formed – planetary embryos – have an elongated shape. This is also observed in the two-dimensional case, but it is even more pronounced in the three-dimensional formulation of the problem. Reducing the exponent in the repulsive and dissipative components allowed this problem to be partially solved (Fig. 8.12). In nonscientific language, the particles are said to become softer as a result of this potential modification, while the potential itself is slightly smoothed. The shape of the forming bodies in this case changes and becomes symmetric in the plane of rotation of the entire system. However, the bodies still remain flattened along the axial symmetry axis of the system. This problem can be resolved by reduction of the dissipation coefficient in the system, which will be discussed below.

Computations showed that the angular and random velocities have strong influence on the number of formed bodies. A series of computations for $2 \cdot 10^4$ particles revealed the domain of values in which the most probable result is the formation of two embryo bodies (Fig. 8.13). This domain is most clearly seen in Fig. 8.14, which plots the dependence of the number of clusters on the initial velocities after processing the data obtained.

A more detailed picture appears as follows. At a low angular velocity, a strong collapse is observed. The particles under the forces of mutual attraction rush toward the system's center, where a single protoplanet is formed. When the angular velocity increases, the gravitational forces are balanced by the centrifugal ones. The system is

Section 8.4   3D-model of evaporative fragmentation

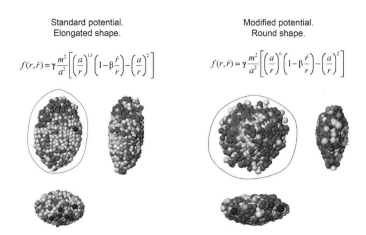

Figure 8.12. Influence of the interaction potential on the shape of the bodies in the plane of rotation of the system.

		v_{rand}/v_s										
		0.60	0.58	0.56	0.54	0.52	0.50	0.48	0.46	0.44	0.42	0.40
ω_0/ω_s	0.65	1.00	1.00	1.00	1.00	1.00	1.00	1.00	1.00	1.00	1.50	1.50
	0.7	1.00	1.00	1.00	1.00	1.00	1.00	1	1.25	1.00	1.50	2.00
	0.75	1.00	1.00	1.00	1.00	1.00	1.00	1.25	1.50	1.25	1.50	2.41
	0.8	1.00	1.00	1.00	1.00	1.00	1.00	1.50	1.50	1.75	2.50	3.00
	0.85	1.00	1.00	1.00	1.00	1.00	1.00	1	1.85	2.14	2.714	2.33
	0.9	1.00	1.00	1.00	1.00	1.00	1.00	2.00	1.28	1.86	2.57	2.75
	0.95	1.00	1.00	1.25	1.25	1.00	2.00	1.25	2.29	2.43	2.57	3.17

Figure 8.13. Mean number of clusters for various initial particle velocity distributions for a system of $2 \cdot 10^4$ particles.

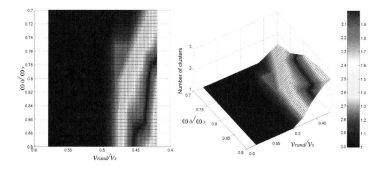

Figure 8.14. Dependence of the number of clusters on the angular and random velocities.

more equilibrated, which makes the formation of two or more bodies possible. As the centrifugal forces grow further, the balance is upset and the model loses its physical meaning. The particles fly apart in different directions and the formation of clusters ceases.

The influence of the initial random particle velocity on the system is also predictable. In its absence, the gravitational instability gives rise to a host of particle clusters. As a consequence, a host of clusters are formed, which subsequently begin to collide with one another and to coalesce into larger ones. Again, if the random velocities are capable of balancing the gravitational interaction, then fewer clusters appear. Accordingly, the number of forming bodies decreases to three, two, or one. If the random motion of the particles in the system is too large, then all clusters are destroyed and no planetary embryo is formed in this case.

Let us note that the optimal velocity values shown in Figs. 8.13–8.14 are slightly different from the values obtained above in the 2D case, where the maximum probability for a two-body formation was realized for $v_{\text{rand}} = 0.68 v_s$ and $\omega_0 \approx 0.76 \omega_s$ (in the 2D case only the angular velocity was optimised, the maximum random velocity was just set to the value $v_{\text{rand}} = 0.68 v_s$).

The optimum values for angular and random velocities obtained for the 3D case satisfy the possible conditions of existence of the gas-dust clouds and they are in good correspondence with the works Vityazev at al. (1990), Marov and Kolesnichenko (2012), Safronov (1969), and the possible values of the angular momentum of the Earth-Moon system, which at that stage of its development was higher than the current angular moment of this system.

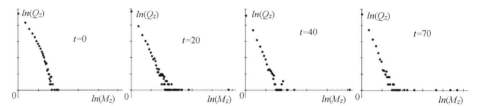

Figure 8.15. Cluster size distribution.

Let us pass from the influence of the initial velocities to analyzing the formation and change of clusters. Figure 8.15 presents the cluster mass distribution on a double logarithmic scale and its change during the computation. Here, we apply the following technique for measuring the cluster size (mass). The space containing particles is partitioned into cubic cells. The particles in the same cell or in the neighboring cells are assumed to belong to one cluster. Subsequently, the dependence of the number of clusters on their sizes is recorded. The linear region of the plots corresponds to the "noise" component – small clusters whose number during the computation oscillates only slightly. The points lying on the horizontal axis rightward of the linear segment characterize the large and easily discernible (in the figures of particle configurations) clusters.

The plots in Fig. 8.15 per se do not give very much information. However, if similar plots for several computations are compared and analyzed, then some trends can be revealed. In Fig. 8.16, the logarithm of the mass of the two largest (in size) clusters is plotted against time. The results obtained in several numerical experiments with identical initial parameters are superimposed. We see that, despite the outward differences in cloud dynamics for various computations, there are common trends for the largest clusters to grow.

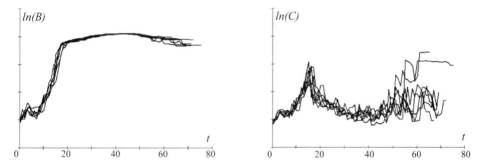

Figure 8.16. Logarithm of the mass of the largest (left) and second largest clusters versus time for several computations with identical initial parameters.

The computations showed that the type of interaction between the bodies as they collide changes with decreasing dissipation. When the dissipation is strong, the bodies are fairly dense and the collision process resembles an impact. In contrast, for a weak dissipation, the bodies are less dense. The particles inside them continue to move randomly and to collide with one another. In this case, soft coalescence of bodies occurs during their collision. In addition, weaker dissipation increases the condensation time for the clusters. As a result, the bodies have a more regular round shape. Thus, a decrease in the dissipation coefficient in combination with a change in the exponent in the repulsive and dissipative components of the interaction force allows the problem of an irregular elongated shape of the embryos to be solved (Fig. 8.17).

Remarkably, a reduction in dissipation affects the details of the process but does not change to any great extent the general trends in the behavior of the system as a whole, such as the mean number of forming bodies or the mean clustering time. Thus, the dissipation coefficient is unlikely to be a parameter capable of slowing down the initial formation of planetary embryos. Regardless of whether this coefficient is large or small, the system fairly rapidly (during several revolutions around its axis) breaks up into separate clusters whose density exceeds that of the remaining dust by several times (Fig. 8.18).

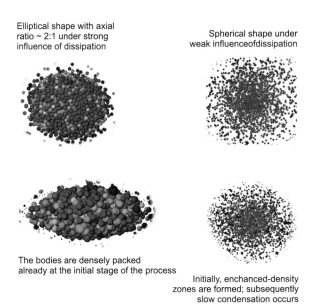

Figure 8.17. Influence of dissipation on the shape of the embryo bodies.

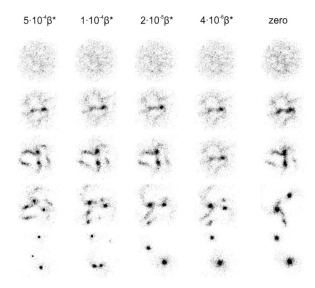

Figure 8.18. Comparison of the dust cloud evolution processes for various dissipations (the dissipation values are shown at the top; each column contains sequential frames of the cloud evolution).

8.4.3 Variation of number of particles

The dust cloud model was composed in such a way that a similarity in the behavior of systems with different numbers of particles could be expected. This is a very important point, because it is just impossible to perform calculations with the number of bodies that are involved in a real process of this kind. No modern computer can cope with such a problem. Therefore, the behavior of a real system can be judged only from similarity considerations.

As our experiments showed, systems with different numbers of particles behave similarly. For almost any computation with a small number of particles a similar (in the pattern of collapse) computation with a number of particles larger by one or two orders of magnitude can be found among all results. Fig. 8.19 gives an example of two computations with different numbers of particles (2×10^4 and 2×10^5) and an approximately similar pattern of evolution of the system. Another instance of similar computation results for a system containing 10^5 particles is shown in Fig. 8.20 (sequential stages) and Fig. 8.21 (top and side view).

However, the computations show that the mean number of forming clusters and the characteristics of their growth depend slightly on the number of particles. For example, the domain of initial parameters providing the most probable formation of two bodies is shifted to the zone of slightly higher random velocities. The new effect that is observed for the larger systems is formations of temporary "bridges" between the bodies being condensed, as in Fig. 8.22 (evolution of the system containing half a million particles with higher angular velocity than in Fig. 8.21). The "bridges" are denser then the distributed dust material but much less dense then the clusters. After some time the "bridges" disassemble in smaller clusters and are accreted by the biggest bodies. Thus, there is similarity for the different orders in the number of particles; however, additional investigations of the scale effects are desirable.

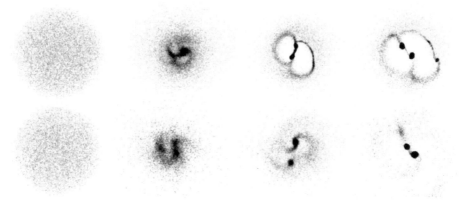

Figure 8.19. An example of computations with different numbers of particles but with similar results, the number of particles is $2 \cdot 10^5$ (top) and $2 \cdot 10^4$ (bottom).

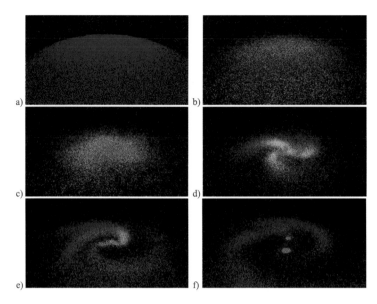

Figure 8.20. Sequential stages of rotational collapse of a cloud of evaporating particles (oblique projection). Number of particles is 10^5 (for details see Part II. Chapter 8).

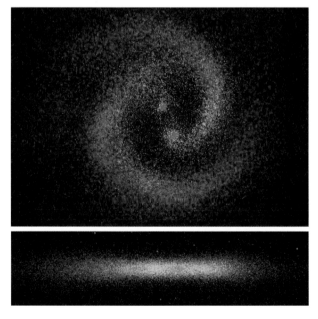

Figure 8.21. Formations of two bodies in a system containing half a million particles: top view (up), side view (bottom).

Section 8.4 3D-model of evaporative fragmentation   123

Figure 8.22. Formation of temporary "bridges" between the bodies being condensed ($5 \cdot 10^5$ particles).

8.4.4 General trends in the system behavior

Thus, having analyzed all of the available results, we can draw the following conclusions (Fig. 8.23). First, the assumption about the similarity of the problem in the two-dimensional formulation to its three-dimensional analog turned out to be, in general, correct. The problems that emerged when passing to the three-dimensional model were successfully resolved after slight changes in the system's parameters and initial conditions. The initial velocity distribution, namely the initial angular velocity and initial random velocity of the particles, as expected, affects strongly the number of ultimately forming bodies. An increase in the initial velocities leads to rapid destruction of the system immediately after the start of computations. In contrast, their decrease leads to collapse and rapid formation of a large number of clusters. The dissipation coefficient has a rather weak effect on the result. Therefore, as has already been stated, it is unlikely to be a factor capable of slowing down the process. However, for example, the as yet disregarded influence of the Sun may produce such an effect.

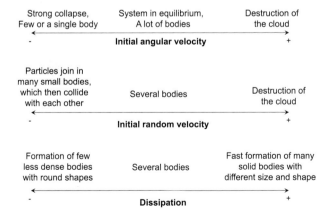

Figure 8.23. General trends in the system's behavior when changing its basic parameters.

Thus on the basis of the developed simulation technology numerous computing experiments for the rotational collapse of the gas-dust cloud system were carried out, in which the influence of the system's initial micro- and macroparameters on the computational results was studied qualitatively. The domain of parameters at which the system demonstrates the behavior of interest – rotational collapse with the formation of a two-body system – was obtained. General trends in the formation of the stable clusters of particles were revealed. To all appearances, further studies will require a complication of the physical model. This will entail the necessity of increasing the speed of computations. Therefore, the computational and physical development must go in parallel.

Chapter 9

Dynamic modeling of accretion

In the previous chapter it was shown that fragmentation is feasible and the main parameters of this process were determined. A solution of the fragmentation problem was achieved using the similarity parameter, which allows consideration of a system of arbitrary mass, while keeping the ratio of the mass and angular momentum close to the known value for the Earth-Moon system. In the case of the Earth and Moon an important mass limitation should be accounted for, which is that the mass of the Moon embryo cannot be more than approximately 0.01 of the total mass of the initial cloud (the final relation being observed nowadays for the Moon and Earth masses is 1 : 81.3). Therefore the quantitative solution of the accretion problem should provide such evolution of the system that the given mass relation for the embryos will give the known masses for the Earth and Moon at the end of the accretion process. One of the main questions to be answered in this chapter is: how will the material be distributed among two embryos, and how will their growth and change of relative sizes occur?

Sections 9.1 and 9.2 of this chapter are based on the papers Vasiliev et al. (2004), Galimov and Krivtsov (2005), Vasiliev et al. (2011).

9.1 Computational model

The two embryos rotate around a common center of mass. Embryo growth occurs at the expense of the fall of material particles from the surrounding space on the embryos, and the fall of particles occurs in the gravitational field of the double system.

Average densities of the condensed embryos and the dust particles are assumed to be equal.

Let us denote masses of embryos by m_1 and m_2 (for definiteness, we will assume $m_1 > m_2$), $m = m_1 + m_2$ is the total mass of the system, a is the distance between embryos, and G is the gravitation constant. Only the mutual attraction of embryos and their gravitational influence on dust particles are taken into account. The gravitational interaction of the dust particles and their effect on embryos are not considered. Radii of the embryo orbits, a_1, a_2 and the angular velocity ω of their rotation around the common center of mass of the system are assumed to be constants defined by formulas

$$a_1 = \frac{m_2}{m}a, \quad a_2 = \frac{m_1}{m}a, \quad \omega^2 = \frac{Gm}{a^3}. \tag{9.1}$$

For simplification, the rotation of the system around the Sun is not taken into account in the given statement of the problem. Since the region under consideration is located within the Hill sphere for the largest of the bodies, such simplification is justified. In our paper, comprehensive numerical-analytical modeling of the process of accumulation of particles by the embryos is carried out. The purpose of this analysis is to find out how the growth of embryos at the expense of particles falling on them will occur. The study is performed on the basis of computing experiments and analytical calculations at different variances of the statement of the problem and its parameters. The problem is considered in the two-dimensional and three-dimensional statement. In all numerical experiments, direct calculations of trajectories of each particle are carried out. These trajectories are rather various and complex: a particle can execute several revolutions around the center of mass of the system until it falls on one of the embryo bodies.

A system of two bodies that rotate in the plane XY in circular orbits around a common center of mass is considered. In the 3D statement, the initial positions of the dust particles are uniformly distributed over the cylindrical surface C_{Rh} with the radius R and height $2h$. The axis of the cylindrical surface C_{Rh} coincides with the common center of the circular orbits of the main bodies. Masses of the particles are equal, and the initial velocities are zero. Coordinates of every subsequent particle are chosen randomly on the surface C_{Rh}, and then its trajectory in the gravitational field of the double system is calculated until it touches the surface of any one of the bodies (the surfaces of the embryos are modeled by spheres with radii r_1 and r_2). The computation is also stopped if a particle flies out beyond the limits of the sphere of the radius R with the center in the center of mass of the system. Then the next particle is brought onto the surface C_{Rh}, and the process is repeated. As a result of the multiple repetition of the calculation, the number of the particles fallen on the first and second body is recorded. In the given statement, this registration of the particles describes the change of the masses of the main bodies. The influence of the particles on the motion of the main bodies is not taken into account. In the two-dimensional statement the only difference is that the particles are brought onto a circle that is the intersection of the cylinder C_{Rh} with the plane XY (Fig. 9.1).

In the statement of the computer experiment described above, the mass change of the planet embryos is not taken into account, and only the number of particles, n_1 and n_2, fallen on the first and second body, respectively, is determined. However, these data allow the construction of the analytical model of the embryos growth. Let us show this. From the computer calculation, the function f that connects the mass ratio m_2/m_1 of the embryos with the proportion of the particles fallen on them, n_2/n_1, can be obtained:

$$\frac{n_2}{n_1} = f\left(\frac{m_2}{m_1}\right), \qquad (9.2)$$

Let us introduce a dimensionless parameter equal to the mass ratio of the embryos

$$\xi = \frac{m_2}{m_1}, \quad 0 < \xi < 1 \qquad (9.3)$$

Section 9.1 Computational model

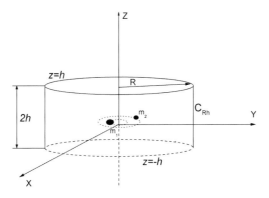

Figure 9.1. Geometrical model of the system.

Obviously, the function $f(\xi)$ has to have the following properties:
$$0 < f(\xi) < 1; \quad f(0) = 0, \quad f(1) = 1. \tag{9.4}$$

Now let us assume that the fall of the particles lasts over a long period of time, considerably exceeding the computer calculation time. Then the masses of the embryos will increase, where the ratio of the mass increase rates of the embryos $\dot{m}_2/\dot{m}_1$ is equal to the ratio n_2/n_1 (assuming that the densities of the material of the embryos and the dust particles are equal). As a result, the following system of equations can be written for the growth of the planet embryos

$$\frac{\dot{m}_2}{\dot{m}_1} = f\left(\frac{m_2}{m_1}\right), \quad m_1 + m_2 = m(t). \tag{9.5}$$

where $m(t)$ is the total mass of the system that is assumed to be a known function of time.

Using the obtained equations, let us study how the mass ratio of embryos ξ will change with time. With this purpose, let us calculate the derivative $\dot{\xi} = (m_2/\dot{m}_1)$ by using Equation (9.5) and expressing the masses of the embryos in the terms of m and ξ:

$$\dot{\xi} = \Phi(\xi)\frac{\dot{m}}{m}, \tag{9.6}$$

where the function $\Phi(\xi)$ is defined by the relation

$$\Phi(\xi) = (1+\xi)\frac{f(\xi) - \xi}{f(\xi) + 1} \equiv \left(\frac{1}{1+\xi} + \frac{1}{f(\xi) - \xi}\right)^{-1}. \tag{9.7}$$

The differential Equation (9.6) is easily reduced to quadratures

$$\int_{\xi_0}^{\xi} \frac{d\varsigma}{\Phi(\varsigma)} = \ln\frac{m(t)}{m_0}, \tag{9.8}$$

where ξ_0 and m_0 are the initial values of corresponding quantities. After calculating the integral (9.8) and finding $\xi(t)$, the masses of the embryos can be calculated by formulas

$$m_1 = \frac{m}{1+\xi}, \quad m_2 = \frac{\xi m}{1+\xi}. \tag{9.9}$$

However, the important conclusions on the dependence behavior $\xi(t)$ can be drawn without calculating the integral (9.8). In particular from Equations (9.6), (9.7) it follows that the sign of $\dot{\xi}$ coincides with the sign of the difference $f(\xi) - \xi$ (all other multipliers are positive, since according to the statement of the problem, the masses of the embryos increase with time). The dynamic equilibrium $f(\xi) = \xi$ (the masses of the embryos increase with time, but their ratio remains invariable) corresponds to the case $\dot{\xi} = 0$. In order to obtain a condition of stability of some equilibrium position, let us vary Equation (9.6) in the neighborhood of the equilibrium point $\xi = \xi_*$ that gives

$$\delta\dot{\xi} = \Phi'(\xi_*)\frac{\dot{m}}{m}\delta\xi, \tag{9.10}$$

where $d\xi$ is the variation (the deviation from the equilibrium position). It follows from the obtained equation that at non zero $\Phi'(\xi_*)$, the necessary and sufficient condition of stability is the condition

$$\Phi'(\xi_*) < 0 \Leftrightarrow f'(\xi_*) < 1. \tag{9.11}$$

The second inequality from Equation (9.11) is obtained using the identity $f(\xi_*) = \xi_*$ For the analysis of the change of the quantity ξ with time, it is convenient to examine dependences $f = f(\xi)$ and $f = \xi$ on the one diagram (Fig. 9.2). If the curve $f = f(\xi)$ is located above the straight line $f = \xi$, then the quantity ξ increases; otherwise it decreases. A point on the curve $f(\xi)$ corresponds to a state of the system; with the course of time the position of this point changes in accordance with the direction of change of ξ. The point of dynamic equilibrium $f = f(\xi)$ corresponds to the intersection of the curve $f = \xi$ with the straight line $\xi = \xi_*$. It is evident from Fig. 9.2 that the equilibrium is stable when, deviating from it in the direction of the increase of ξ, the curve $f = f(\xi)$ finds itself below the straight line $f = \xi$; otherwise, the equilibrium is unstable. Obviously, this condition is equivalent to the stability criterion obtained above (9.11).

Thus, if the function $f(\xi)$ is obtained from the computer modeling, then the evolution of the system can be described analytically using Equations (9.6–9.8) or analyzed graphically on the basis of Fig. 9.2.

According to the analytical consideration performed above, the problem of the numerical modeling is reduced to the determination of the function $f(\xi)$ that connects the mass ratio of bodies with the proportion of particles fallen on them, $n_2 = n_1$. The typical form of the dependence $f(\xi)$ obtained as a result of the computer modeling is presented in Fig. 9.3. It is evident from the figure that in almost the entire range of values it holds that $f(\xi) < \xi$. Consequently, the mass ratio decreases with time – the

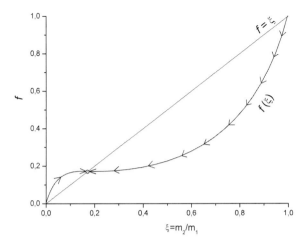

Figure 9.2. Schematic representation of the captured particles ratio as a function of the mass ratio of the embryos.

smallest body accumulates considerably less dust particles than the largest one (not in proportion to their sizes), and, as a result, its relative mass decreases. However, this process cannot last to infinity, at some small value $\xi = \xi_*$ the curve $f = f(\xi)$ intersects the straight line $f = \xi$, and dynamic equilibrium comes: masses increase, and the proportion between them does not change. At the same time $f'(\xi) < 0$, i.e., the situation is analogous to the situation depicted in Fig. 9.2: the equilibrium is stable. This intersection is shown in greater detail in Fig. 9.4; one can see that equilibrium comes at $\xi_* = 0.05$.

Thus, under the unlimited supply of mass to the system of rotating bodies, the ratio of their masses will tend to a small but still fixed value. However, the system under

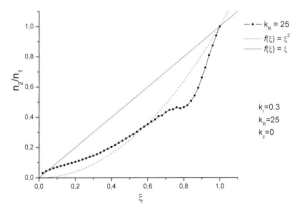

Figure 9.3. Typical form of the dependence $f(\xi)$ obtained as a result of computer modeling.

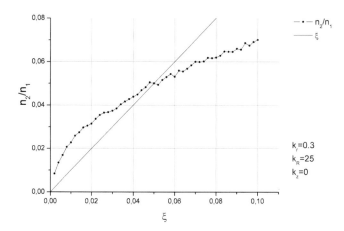

Figure 9.4. Instance of a small ratio of embryo masses.

consideration contains three significant dimensionless parameters affecting the result of the computations

$$k_r = \frac{r_1}{a}, \quad k_R = \frac{R}{a}, \quad k_z = \frac{h}{R}, \qquad (9.12)$$

where a is the distance between the bodies, r_1 is the radius of the biggest body, and h and R are the height and radius of the cylinder C_{Rh}, on which the initial positions of the dust particles are set. Dimensionless parameter k_r characterizes relative remoteness of the embryos from each other, parameter k_R characterizes relative remoteness of the initial position of the particles, and parameter k_z characterizes relative thickness of the initial cloud. At $k_z = 0$, a three-dimensional statement is equivalent to a two-dimensional one. Results, presented in Fig. 9.3, correspond to the following values of parameters

$$k_r = 0.3, \quad k_R = 25, \quad k_z = 0, \qquad (9.13)$$

In this and all subsequent computer experiments, the amount of the falling particles is $n = 10^5$ that allows obtaining sufficiently smooth graphs of the dependence $f(\xi)$, and the further increase of the particle number has little influence on the result.

Let us analyze the influence of dimensionless parameters on results of the calculation. The parameter k_R is the remoteness of the initial position of particles. In Fig. 9.5, a convergence of the results under the increase of k_R is studied. Values of other parameters are: $k_r = 0.3$, $k_z = 0$. Calculations have shown that the convergence of functional dependence occurs approximately at $k_R = 25$; the further increase of this parameter has little influence on the result. Thus, this is the maximum value of this parameter to be used in the computations.

The parameter k_r is the remoteness of embryos from each other. In contrast to k_R, the parameter k_r characterizes the physical side of the problem, rather than the computational one. Fig. 9.6 shows results of this parameter variation at $k_R = 25$, $k_z = 0$.

Section 9.1 Computational model

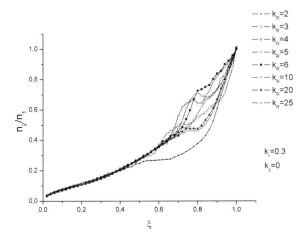

Figure 9.5. Variation of the parameter k_R characterizing relative size of the feeding area.

One can see that the smoothest dependence $f(\xi)$ is realized at $k_r = 0.45$, i.e., when embryos are initially at a short distance from each other. This is possibly connected with the decrease of the probability of flight of falling particles between embryos. As k_r decreases, the dependence $f(\xi)$ becomes less regular, and oscillations appear. In Fig. 9.6 the graph of the ratio of the ejected particle number n_3 to the total number of fallen particles n is also shown.

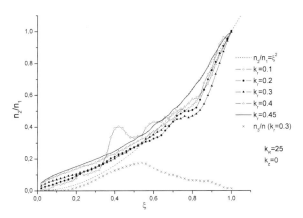

Figure 9.6. Variation of the parameter k_r characterizing the size of the embryos relative to the distance between them.

The value $k_r = 0.3$ was chosen as the base value for the majority of further calculations; this allows us to obtain the rather regular dependence $f(\xi)$ and is in good agreement with results of the previous stage of studies on the modeling of double system formation in the process of the rotational collapse of the gas-dust cloud (see Chapter 8).

The parameter k_z is the relative thickness of the particle cloud. Figure 9.7 shows the comparison of results obtained for different 3D experiments ($k_z > 0$) and 2D experiment ($k_z = 0$). The values of the other parameters are $k_R = 25$, $k_r = 0.3$. The figure confirms the similarity between 3D and 2D statements. However, more detailed consideration reveals that for small values of ξ the influence of the parameter k_z is essential. It is evident from Fig. 9.8 that as the parameter k_z increases, the point of intersection of the curve $f = f(\xi)$ with the straight line $f = \xi$ moves to the left along the axis. In other words, for the 3D case the equilibrium state is reached at smaller ratios m_2/m_1 then for the 2D case. When the parameter k_z increases further, this intersection disappears so that the embryos' mass ratio will tend to zero under the unlimited mass supply to the system.

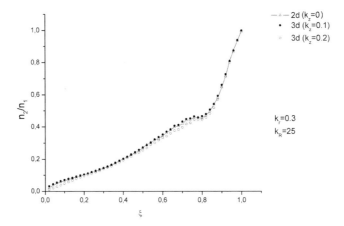

Figure 9.7. Comparison of the results of 2D and 3D simulations.

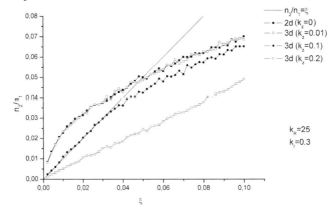

Figure 9.8. Comparison of the results of 2D and 3D simulations for the small mass ratios.

Figure 9.9 shows the change of dependence $f(\xi)$ under the considerable increase of the parameter k_z. From the figure it follows that as the cloud thickness increases, the functional dependence keeps its general form, but it becomes more monotonic allowing a power function approximation.

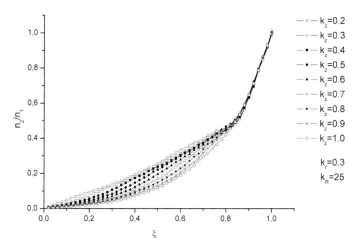

Figure 9.9. Variation of the parameter k_z characterizing the relative thickness of the particle cloud.

Thus the numerical experiment has shown that under the variation of dimensionless parameters in the wide range, the dependence $f = f(\xi)$ remains smooth and lies below the straight line $f = \xi$, at least for not too small values of ξ. Hence, as the accumulation of the dust particles continues, an unproportional growth of the embryos occurs – the ratio of their masses $\xi = m_2/m_1$ decreases. This means that the greater mass grows much faster than the smaller one, attracting more of the dust material. When a sufficiently small value of ξ is reached then dynamic equilibrium (the proportional growth of embryos) can occur.

The consideration of results of the numerical experiment shows that in many cases (Figs. 9.8–9.9) the function $f(\xi)$ can be sufficiently well approximated by a power law

$$f = \xi^{k+1}; \quad k > 0. \tag{9.14}$$

Then the first equation of the system (9.5) is easily integrated, and this leads to the following system of algebraic equations:

$$m_2^{-k} - m_1^{-k} = c^{-k}, \quad m_1 + m_2 = m(t), \tag{9.15}$$

where c is the integration constant that has the dimension of mass. The obtained system implicitly specifies the desired functions $m_1(t)$ and $m_2(t)$. If the supply of mass is unlimited, then $m(t) \to \infty$, and consequently the mass of the biggest body tends to infinity: $m_1(t) \to \infty$. But then we obtain from the first equation of the system (9.15) that the smallest mass tends toward a constant as follows:

$$m(t) \to \infty \Rightarrow m_2(t) \to c = (m_{20}^{-k} - m_{10}^{-k})^{-\frac{1}{k}}. \tag{9.16}$$

where m_{10} and m_{20} are masses of embryos at the initial time. Thus, under the unlimited supply of mass, the biggest body accumulates almost all mass of the cloud, whereas sizes of the smallest body remain to be limited. The same conclusions can be obtained by calculating the integral in equation (9.8); this leads to the following implicit dependence of ξ on t:

$$\frac{1+\xi}{\xi} \sqrt[k]{1-\xi^k} = \frac{1}{c}m(t). \tag{9.17}$$

Under the unlimited supply of mass, the equation (9.17) gives

$$m(t) \to \infty \Rightarrow \xi \to \frac{m(t)}{c}, \tag{9.18}$$

which agrees with the conclusions obtained above.

Let us consider in more details the case $k = 1$ that corresponds to the quadratic function $f(\xi) = \xi^2$. Figure 9.6 shows how much the quadratic function agrees with the numerical results for 2D simulations. The approximation is appropriate for not too small values of ξ. At small ξ, the dependence $f = f(\xi)$ can deviate from the power dependence and intersect the straight line $f = \xi$ (according to results of computer experiments presented above). As the system approaches this point, the ratio m_2/m_1 stops changing, and later masses m_1 and m_2 go on growing proportionally to each other. This effect is less pronounced in the 3D case (Fig. 9.9), where for sufficiently thick clouds the best correspondence with the quadratic law can be observed. For the quadratic function $f(\xi) = \xi^2$ the equation (9.17) can be solved explicitly:

$$\xi(t) = \frac{1}{2}\left(\sqrt{1 + 4\frac{c^2}{m^2(t)}} - 1\right)\frac{m(t)}{c}, \quad c = \frac{m_{10} m_{20}}{m_{10} - m_{20}}. \tag{9.19}$$

After finding $\xi(t)$, masses of embryos $m_1(t)$ and $m_2(t)$ are calculated by formulas (9.9):

$$m_1(t) = \frac{m(t)}{1+\xi}, \quad m_2(t) = \frac{\xi m(t)}{1+\xi}.$$

An important case in practical terms occurs when the known parameters are the values of the masses at the final moment of time, which corresponds to the present day state of the system. Let us note for simplicity these values as M_1, M_2, and $M = M_1 + M_2$, respectively. Then formula (9.19) can be represented in the form

$$\frac{m_2}{m_1} = \frac{1}{2\eta}\left(\sqrt{1 + 4\eta^2} - 1\right), \tag{9.20}$$

where m_1 and m_2 are the masses of the Earth and Moon embryos; the dimensionless coefficient ε is defined as

$$\eta = \frac{M}{m}\left(\frac{M_1}{M_2} - \frac{M_2}{M_1}\right)^{-1}, \qquad (9.21)$$

where $m = m_1 + m_2$. If the ratio M/m is known from some source then the formulas (9.20) and (9.21) allow determination of the mass ratio for the Earth and Moon embryos. After this ratio is found the embryos' mass increase in the accumulation process can be calculated as

$$\frac{M_1}{m_1} = \frac{M}{m}\frac{1+m_2/m_1}{1+M_2/M_1}, \quad \frac{M_2}{m_2} = \frac{M}{m}\frac{1+m_2/m_1}{m_2/m_1}\frac{M_2/M_1}{1+M_2/M_1}. \qquad (9.22)$$

9.2 Determination of sizes of Earth and Moon embryos

Let us use the results of the previous section to estimate the numerical values of the embryo masses. The case $k = 1$ of the quadratic function $f(\xi) = \xi^2$ will be used below as the simplest way of getting the appropriate correspondence in most cases. From Equation (9.15) it follows

$$\frac{1}{m_2} - \frac{1}{m_1} = \frac{1}{M_2} - \frac{1}{M_1}, \qquad (9.23)$$

where m_1 and m_2 are the masses of the Earth and Moon embryos, M_1 and M_2 are the current Earth and Moon masses. Let us use dimensionless parameters

$$\xi = \frac{m_2}{m_1}, \quad \varepsilon = \frac{M_2}{M_1}. \qquad (9.24)$$

Multiplication of Equation (9.23) by m_2 after some transformations gives

$$\frac{m_1}{M_1} = \frac{1-\xi\varepsilon}{1-\varepsilon\xi}, \quad \frac{m_2}{M_2} = \frac{1-\xi}{1-\varepsilon}, \quad \frac{m}{M} = \frac{1-\xi^2\varepsilon}{1-\varepsilon^2\xi}. \qquad (9.25)$$

Here $m = m_1 + m_2$, $M = M_1 + M_2$. For the Earth-Moon system $M_2 \ll M_1$. Then considering $\varepsilon = M_2/M_1 \approx 1/80$ as a small parameter we obtain in the first approximation

$$\frac{m_1}{M_1} = \frac{1}{80}\frac{1-\xi}{\xi}, \quad \frac{m_2}{M_2} = 1-\xi. \qquad (9.26)$$

For geochemical reasons the ratio m_2/M_2 should be relatively close to unity, since the Moon preserved its high-temperature embryo composition. On the other hand the ratio m_1/M_1 should be a small value, since the Earth accumulated the most of the cold mass of the cloud and has increased its mass significantly compared to its original embryo. This gives two side constraints on the ξ value – it should be relatively small compared to 1 and big compared to $M_2/M_1 \approx 1/80$. Analysis of Figs. 8.7–8.8 shows that the probable ratio of the embryo sizes is 1/4, which fully satisfies these conditions. In this case according to Equation (9.26) the mass of the Moon increases by approximately 30 % compared to its embryo value, while the Earth embryo mass increases $80/3 \approx 27$ times. These values correspond well with geochemical data. Using (9.25) the total mass increase in the first approximation can be calculated as

$$\frac{m}{M} = \frac{1}{80}\frac{1-\xi^2}{\xi} \approx \frac{1}{80\xi} \qquad (9.27)$$

Using $\xi \approx 1/4$ gives for the total mass an increase value of $80/4 \approx 20$ that can be used to obtain the appropriate temperature estimations (see Section 9.4).

An example of exact calculations using exact formulas (9.25) is shown in Table 9.1. From the table it follows that the approximate calculations gave a very good estimation: according to the exact formulas for the mass ratio $m_1/m_2 = 4.07$ the Earth has increased its mass 26 times while the Moon mass increase is only 31 %.

Table 9.1. Caption missing

Quantity	Symbol	Value	Units
Actual Earth mass	M_1	$5.98 \cdot 10^{24}$	kg
Actual Moon mass	M_2	$7.35 \cdot 10^{22}$	kg
Actual mass ratio	M_1/M_2	81.3	–
Initial mass fraction	m/M	0.047	–
Embryos mass ratio	m_2/m_1	4.07	–
Earth embryo mass	M_1	$2.28 \cdot 10^{23}$	kg
Moon embryo mass	M_2	$5.61 \cdot 10^{22}$	kg
Earth mass increase	M_1/m_1	26.2	–
Moon mass increase	M_2/m_2	1.31	–

In this chapter the problem of growth of the system of two embryos that rotate around the common center of mass and accumulate material from the dust cloud surrounding these embryos has been studied. The problem was considered in a two-dimensional and three-dimensional statement, and it is shown that given a sufficiently thin dust cloud, the results of the two-dimensional and three-dimensional modeling are almost identical. It is discovered that in the two-dimensional statement the following scenario of the bodies' growth is realized under an unlimited supply of mass to the system: both embryos grow boundlessly, and the ratio of their masses tends toward the value at which the smaller body is about 5 % of the mass of the bigger body. In the three-dimensional statement, this scenario is also realized with a thin dust cloud, but the equilibrium ratio of the embryo masses decreases with increased cloud thickness. If the ratio of the cloud thickness to its diameter exceeds the critical value approximately equal to 0.1, then the scenario changes: the bigger body grows boundlessly, the smaller one grows up to a certain limit, and the mass ratio of the embryos tends to zero. An analytical approximation for the numerical dependence of the ratio of embryo growth rates on the ratio of their masses was proposed, which allowed obtaining an analytical solution of the problem.

An analogous problem to embryo growth was considered by V. S. Safronov (2002), where results similar to those discussed above were obtained from an analytical study. Formally, another problem was being considered in Safronov's paper, where the planets rotate around the Sun, whereas in the current consideration they rotate around the common center of mass. However, in both problems the rotation of the system occurs within a feeding medium around the center of a feeding cloud, and this makes it possible to compare them. It was shown (Safronov, 2002) that as two bodies grow in the feeding medium, the mass difference will increase with time, i.e., the bigger body becomes still bigger, and the bigger body grows faster both absolutely and relatively; i.e., the ratio m_2/m_1 rises. This is exactly the same conclusion as was obtained above. The numerical value of the mass ratio according to Safronov can rise to the value 10^{-3}, which is in agreement with the results obtained above for the relatively thin feeding cloud.

The present-day mass relation for the Earth-Moon system (1/80) can be explained either by the thickness of the feeding cloud or by the fact that the supply of mass stopped for some reason. For other planets of the solar system that have considerably smaller ratios of satellite planet mass, the situation can be explained by the longer supply of mass to the system of rotating embryos of these planetary systems.

The main conclusion of this dynamical consideration, which is of primary interest for the geochemical analysis, is that a random initial difference of masses leads to the situation where the smaller body does not significantly gain in mass, while the bigger one accumulates most of the initial particle pool. Thus the Earth embryo collected the major fraction of the surrounding material, whereby the composition of the Earth came closer to the composition of the cloud as a whole than the composition of the Moon, which remains close to the original high temperature composition depleted in volatiles and iron.

9.3 Consideration of precollapse evolution of the gas-dust cloud

It follows from Table 9.1 that the mass of condensed matter in both fragments (the Earth embryo and the Moon embryo) comprises 4.7 % of the total mass of the initial cloud. The latter is assumed to be equal to the sum total of the mass of Earth and Moon. It means that the initial density distribution in the cloud was such that fragmentation occurred when its central dense part had got about 5 % of the total mass during the contraction of the cloud. Let us note that in the modeling performed in Chapter 8 the obtained sizes of the embryos were greater then 1/20 of the size of the cloud. This can be explained by the choice of the initial configuration of the cloud, which for the

3D case was taken as an elliptical cloud with uniform particle distribution. Obviously this approximation is only good for the denser central part of the cloud. The real cloud should have a density that gradually decreases from the center. In other words the central part of the cloud where the fragmentation occurs is surrounded in reality by an area with much lower density, which becomes a feeding pool for the embryos growth up to their current sizes.

In this and the previous chapter a dynamical analysis of the formation of the Earth-Moon system and an investigation of the related physical and computational problems was presented. The obtained results explain from a dynamical point of view the formation of both Earth and Moon from a common source (Chapter 8) and the differences in the current composition of the Earth and Moon (Chapter 9). However, the formation and precollapse evolution of the common source of the system – the dust-gas cloud that gave birth to the Earth and Moon embryos – was not discussed in detail. This stage of the system's evolution requires further investigation. Below only the key points of this process will be introduced and discussed.

We presume that apart from the solid body accumulation process, a part of the solar nebular material that was still in a gas-dust state decomposed to local clumps, which then assembled in large, but for the solar system rather compact gas-dust condensations. Different aspects of this process have been analyzed by many authors (Gurevich and Lebedinskii, 1950; Eneev and Kozlov, 1977; E. M. Galimov, 1995; Marov and Kolesnichenko, 2012). In time the gas-dust condensations formed a planetary-mass gas-dust cloud. This cloud stayed in a quasi-equilibrium state for a period of about 50 million years, during which a weak loss of gas (vapor) and occasional accumulation of smaller gas-dust clouds and planetesimals led to a slow build-up of the cloud. When the main sources of vapor were depleted the cloud began to collapse under its own gravitation, initiating the fragmentation process that was studied in detail in Chapter 8.

The feasibility of dynamic stability of the gas-dust cloud is based on the following aspects.

1. The gas drag repulsion caused by evaporation of the dust particles satisfies the reverse square law, just as gravitational attraction does. This allows complete mutual compensation of these forces. However, there is one important difference between them: the screening that accompanies gas repulsion does not occur with gravitational attraction. For sufficiently low density systems screening can be neglected, for more condensed systems it can decrease the gas repulsion, but at higher evaporation rates the balance between gravitational attraction and gas repulsion can be still be realized.

2. The gravitational force is proportional to the 6th power of the particle radius, gas repulsion is proportional to the 4th power. Therefore for sufficiently small particles equilibrium can always be realized.

3. The intensity of screening in the scale of the whole cloud is reversely proportional to the particle radius (for the constant mass and volume of the cloud). Thus for small particles the screening is higher.

4. In the case of absolutely inelastic collisions of gas molecules with dust particles and absence of gas leakage from the cloud surface the equilibrium of the cloud can be preserved infinitely.

5. In the case of strong screening (in a relatively condensed cloud) the gas pressure forces are short-ranged and do not depend on the size of the particles and their number in the unit volume, but these forces significantly increase with temperature.

6. The equilibrium of the gas phase of the gas-dust cloud can be approximately described on the basis of the Lane-Emden equation for the self-gravitating spherically symmetric polytropic gas cloud (Horedt, 2004). A stable equilibrium for such a cloud can be realized for levels less than 3 on the polytropic index, which corresponds to the temperature increase towards the cloud center. In this case the cloud has a well-pronounced boundary (in contrast with infinite isothermal cloud), which allows localization of the cloud inside the Hill sphere.

7. The rise of temperature with depth of the cloud can be sustained by slow subsidence of the dust particles. Thus for the prolonged existence of the cloud feeding by occasional accumulation of smaller gas-dust clouds and planetesimals is needed. Feeding can also compensate gas leak from the cloud surface.

9.4 Temperature evolution

Assume that the energy loss under the effect of the dissipative component of the interaction force between the particles is transformed into the particles inner energy according to the following equation:

$$\dot{U}_k = \sum_{n=1}^{N} Q(r_{kn}, \dot{r}_{kn}) - \lambda U_k, \quad k = 1, 2, ..., N; \quad (9.28)$$

where U_k is the inner energy of the kth particle; $Q(r, \dot{r})$ is the quantity of heat transformed into the inner energy under the effect of dissipative forces; λ is the coefficient describing the losses of thermal energy due to its transfer from the particle to the surrounding gas, radiation, and gas evaporation from the particle surface. The quantity of heat $Q(r, \dot{r})$ according to Equations (8.2) and (8.3) can be expressed as

$$Q(r, \dot{r}) = -\frac{A_3}{r^{p+1}} \dot{r}^2 = -\frac{\beta \gamma m^2 a^{p-2}}{r^{p+1}} \dot{r}^2. \quad (9.29)$$

The calculation also takes into account the inter-particle heat exchange. The particle temperature can, for a first approximation, be taken as proportional to its inner energy calculated according to Equation (8.21). Note that the influence of thermal effects on the system dynamics is not taken into account.

Let us analyze the temperature effects for the 2D computations of the rotational collapse (Section 8.2.6). Assume that the initial temperature of the cloud is 300 K, and the average heat capacity of the solids is 800 J/(kg K). Application of the above equations gives the time-temperature profile for each particle in the cloud. The level of temperature depends strongly on the mass involved in the collapse process. An estimation of this mass based on dynamical principles and consistent with geochemical arguments (see the next chapter) gives approximately 1/20 of the total mass of the cloud (the current mass of the Earth-Moon system). The corresponding configurations with temperature distributions are depicted in Fig. 6.10 for sequential moments of time. In the process of the rotational collapse two high-temperature embryos of the Earth and Moon are formed. The embryos are surrounded by a distribution of colder dust material. Figure 6.9 shows the temperature profiles for each embryo and the dust. Measurement of the temperatures of the separate clusters or bodies requires special algorithms for computational determination of such objects, which are quite complicated and do not always allow obtaining unambiguous results. The results depicted in Figs. 6.9, 6.10, and 9.10 were obtained on the basis of percolation algorithms (Mukhin and Volkovets, 2004).

From the figures it follows that the average temperature of the dust does not exceed 750 K, which means that there is no significant loss of volatiles from the dust

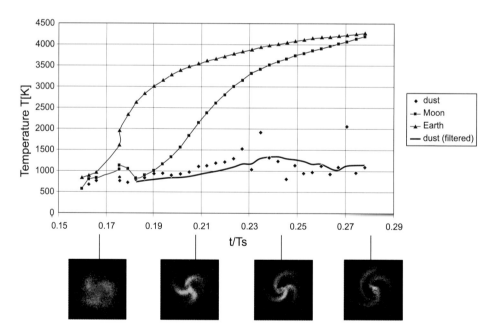

Figure 9.10. Dependence of average temperatures on the collapse time obtained by 3D computation.

component. At the stage of formation of the two dust concentrations (Fig. 6.10) their temperature does not exceed 2000 K, which is sufficient for loss of volatiles and Fe, but the refractory elements are preserved. At the stage of formation of the solid bodies (Fig. 6.10) their average temperature goes up to 4,000 K and more; however the temperature of their surface areas is still not above 2,000 K, so the refractory elements are still not lost.

With this scenario a successful correlation between the dynamical calculations and the geochemical data of the Earth and Moon composition can be obtained. The geochemical aspects of this process are analyzed in more detail in Chapter 6.

The same technique as was described above for the 2D case was used to obtain temperature distributions in the 3D case. The special percolation algorithms for determination of particle concentrations and measurement of their characteristics were developed, including algorithms allowing separate measuring of temperatures at the surface and the core of the clusters. The obtained temperature profiles for the Earth and Moon embryos and for the surrounding dust are shown in Fig. 9.10 (3D computation using 20,000 particles). In general the 3D computations have confirmed the results obtained by the 2D computations.

Conclusions

In this book we have attempted above all to indicate those features of the Moon which constitute a system of constraints on its origin. The Moon and the Earth reveal exceptional similarity in their isotope characteristics, and significant differences in chemical composition.

For several decades the giant impact theory seemed to be the best solution to the Moon origin problem. It was well-grounded dynamically and provided plausible explanations of the major facts, including the deficit of iron in the Moon and excessive angular momentum of the Earth-Moon system. The main difficulty of this hypothesis stems from the fact that the Moon, according to this theory, originates from material of an impactor alien to the Earth. It is difficult to reconcile this with the observed isotopic identity of the Earth and the Moon, which undoubtedly testifies to their origin from a common source. The idea of subsequent homogenization has also proved inadequate to simultaneously explain the coincidence in isotopic identity (including Ti and W isotope composition) and the chemical differences (in refractory and volatile elements) of the Earth and Moon.

The model proposed here is an alternative to the giant impact hypothesis and overcomes most of its major difficulties. The hypothesis suggests formation of the Earth and the Moon from the initial state of a large gas-particle cloud. Its contraction resulted in an adiabatic temperature increase in its interior parts and evaporation of volatile components, including iron, from the surface of particles. At a certain stage the internal part of the dispersed formation underwent fragmentation.

Modeling using particle dynamics shows the feasibility of the fragmentation process if the evaporation of particles is taken into account. Fragmentation lead to the formation of two condensed bodies – the Earth and Moon embryos. A wide range of values for the system parameters has been analyzed and the area of the most appropriate values in the parameter space have been found. After fragmentation the remaining part of the surrounding material was accreted mainly to the Earth. The dynamic analysis shows that a random initial difference of the masses leads to the situation where the smaller body mass does not significantly increase, while the mass of the bigger one increases many times over. Thus the Earth embryo collected the major fraction of the surrounding material, whereby the composition of the Earth came closer to the composition of the cloud as a whole than to the composition of the Moon, which remained close to the original high temperature composition depleted in volatiles and iron.

During this process, a vapor dominated by primary hydrogen (then water) was squeezed out of the cloud. The vapor was expelled from the interparticle space together with the carrier gas (hydrogen and water). This hydrodynamic flow resulted in the loss of volatiles, including Rb, Xe, and Pb, which is reflected in the Rb–Sr, Xe–I–Pu, and U–Pb isotopic systems. Gas-dust accretion was completed at $\sim 120\,\text{Ma}$ after the beginning of the solar system.

What problems does the hypothesis solve, and what facts can it explain?

The existence of a common origin for the Moon and the Earth solves the problem of the coincidence of the ^{16}O–^{17}O–^{18}O fractionation lines for the Earth and the Moon, as well as the problem of their identical Ti, W, Cr, Mg, Si, Li isotopic compositions. The proposed hypothesis explains the depletion of the Moon in Fe. It is shown that the loss of Fe is correlated with the enrichment of the Moon in refractory elements (Al, Ca, and Ti). Isotopic exchange in the vapor-particle system explains why volatile loss from the Moon was not accompanied by isotope fractionation. The dependence of the probability of fragmentation on the angular momentum of the gas-dust disk explains why the Earth has a satellite, while the other inner planets do not.

The origin of a planet from initial gas-dust accumulation implies the long existence of an extended envelope consisting of particles and vapor. The hydrodynamic regime imposed by an ascending hydrogen (water) flow (hydrodynamic escape) solves the problem of the loss of a range of volatile elements, including Rb, Xe, and Pb, from the Earth during the first $\sim 120\,\text{Ma}$. In previous work the loss of heavy volatiles was attributed to various processes: catastrophic impact-related removal of the primordial atmosphere (in the case of Xe), Pb migration into the core, etc. However, each of these explanations contradicted certain observations. The present model provides consistent explanations for these observations. It also provides a solution to the long-standing problem of the formation of a metallic core from the initial FeO state, which required the removal of a tremendous equivalent amount of oxygen from the Earth.

The Hf–W, Rb–Sr, Xe–I–Pu, and U–Pb isotopic systems provide constraints on the main events of the evolution of the proto-Earth-Moon-vapor-particle body: fragmentation with formation of the Moon and Earth embryos occurred about 50 Ma, and full accomplishment of accretion about 120 Ma after the beginning of the solar system. Hence the age of the Earth as a consolidated planet is $\sim 4.44\,\text{Ga}$ rather than the customary $4.56\,\text{Ga}$.

We realize that many aspects of the origin of the Earth-Moon system have remained untouched. If the proposed hypothesis is correct in principle, the mechanism of planet formation and interpretation of many important geochemical relationships need to be fundamentally revised. We believe that the present considerations give cause for this undertaking.

Acknowledgments

We thank N. V. Bondar, V. E. Kulikovski, A. A. Le-Zaharov, M. A. Mukhin, O. K. Timonina, S. V. Vasiliev, and I. B. Volkovets for their help in different stages of this work.

References

Adams J. C., On the secular variation of the eccentricity and inclination of Moon's orbit. *Mon. Not. Roy. Astr. Soc.* **19**: 206–208, 1859.

Agee C. B., Li J., Shannon M. C. and Circone S., Pressure-temperature phase diagram for the Allende meteorite. *Journal of Geophysical Research* **100**(B9) 1995. doi: 10.1029/95JB00049. issn: 0148-0227.

Akim E. L., Determination of the gravitational field of the Moon from the motion of the artificial lunar satellite "Lunar-10". *Dokl. Akad. Nauk* SSSR **170**: 799–802, 1966.

Albarede F., Volatile accretion history of the terrestrial planets and dynamic implications. *Nature* **461**: 1227–1233, 2009

Alibert C., Norman M. D. and McCulloch M. T., An ancient Sm-Nd age for a ferroan noritic anorthosite clast from lunar breccia 67016. *Geochim. Cosmochim. Acta* **58**: 2921–2926, 1994.

Allègre C. J., Manhes G. and Göpel C., The Age of the Earth, *Geochim. Cosmochim. Acta* **59**: 1445–1456, 1995.

Allègre C. J., Manhes G. and Göpel C., The Major Differentiation of the Earth at 4.45 Ga. *Earth Planet. Sci. Lett.* **267**: 368–398, 2008.

Acta **53**, 197–214, 1989.

Altshuler L. V., Krupnikov K. K. and Brazhnic M. I., Dynamical compressibility of metals for pressures from four hundred to four million atmospheres. *J. Experimental & Theoretical Phys.* **34**: 886–893 (in Russian), 1958.

Amelin Y., Krot A. N., Hutcheon I. D. and Ulyanov A. A., Pb isotopic ages of chondrules and Ca, Al-rich inclusions. *Science*, **297**: 1678–1683, 2002.

Anders E. and Grevesse N., Abundances of the Elements: Meteoritic and Solar, *Geochim. Cosmochim Acta* **53**: 197–214, 1989.

Ariskin A. A., Parental magmas of lunar troctolites: Genetic problems and estimated original compositions. *Geochemistry International*, **45**(5): 413–427, 2007.

Araki H., Tazawa S. and Noda H., et al., Lunar global shape and polar topography derived from Kaguya-LALT laser altimetry. *Science*, **323**: 897–900, 2009.

Arkani-Hamed J., Delayed activation of Martian core dynamo. In: Proc. 43rd Lunar and Planet. Sci. Confer., abst. 1563, 2012.

Arnold J. R., Ice in the lunar polar regions. *Jour. Geophys. Res*, **84**: 5659–5668, 1979.

Bagdassarov N., Solferino G., Golabek G. J. and Schmidt M. W., Centrifuge assisted percolation of Fe-S melts in partially molten peridotite: Time constraints for planetary core formation. *Earth Planet. Sci. Lett.* **288**: 84–95, 2009.

Ball R. S., On the accurate of great tide since the commencement of the geological epoch. *Nature* **27**: 201–203, 1882.

Ballhaus C. and Ellis D. J., Mobility of core melts during Earth's accretion. *Earth Planet. Sci. Lett.* **143**: 137–145, 1996.

Barnes J. and Hut P. A., A hierarchical O ($N \log N$) force calculation algorithm. *Nature* **324**: 446–449, 1986.

Basilevsky A. T., On the evolution rate of small craters. In: Proc 7th Lunar Sci Conf., **1**: 1005–1020, 1976.

Becker H. et al., Highly siderophile element composition of the Earth's primitive upper mantle: Constraints from new data on peridotite massifs and xenoliths. *Geochim. Cosmochim. Acta* **70**: 4528–4550, 2006.

Belbruno E. and Gott J. R., Where did the Moon come from? *Astronom J.* **129**: 1724–1745, 2005.

Belleman R. G., Bedorf J. and Zwart S. F. P., High performance direct gravitational N-body simulations on graphics processing units II: An implementation in CUDA. *New Astronomy* **13**:103–112, 2008.

Benz W., Slattery W. L. and Cameron A. G. W., The origin of the Moon and single impact hypothesis I. *Icarus* **66**: 515–535, 1986.

Benz W. and Cameron A. G. W., Terrestrial Effects of the Giant Impact, *Origin of the Earth*, Eds.: H. E. Newsom and J. H. Jones, 61–67,Oxford Univ. Press, New York, 1990.

Boyce J. W. et al., Lunar apatite with terrestrial volatile abundances. *Nature* **466**: 466–469, 2010.

Boss A. P. and Peale S. J., Dynamical constraints of the origin of the Moon. In: *Origin of the Moon*, Eds.: W. K. Hartmann, R. J. Phillips and G. J. Taylor: 59–101, Lunar and Planetary Institute, Houston, 1986.

Borg L. E., Norman M. D., Nyquist L. E., Bogard D. D., Snyder G. H., Taylor L. A. and Lindstrom M. M., Isotopic studies of ferroan anorthosite 62236: A young lunar crustal rock from a light rare-earth-element-depleted source. *Geochim. Cosmochim. Acta* **63**: 2679–2691, 1999.

Bouvier A., Wadhwa M. and Janney P., Pb-Pb isotope systematics in an Allende chondrule. *Goldschmidt Conference Abstracts* July 13-18, Vancouver, Canada, A104, 2008

Bryden G., Chen X., Lin D. N. C., Nelson R. P. and Papaloizou J. C. B., Tidally Induced Gap Formation in Protostellar Disks: Gap Clearing and Suppression of Protoplanetary Growth. *Astrophys. J.* **514**: 344, 1999.

Buchachenko A. L., Galimov E. M. and Ershov V. V., Isotope effect induced by spin-spin interaction in chemical reactions. *Dokl. Acad. Sci.* USSR **228(2)**: 1976.

Buck R. and Toksoz M. N., The Bulk Composition of the Moon Based on Geophysical Constrains, in Proceedings of 11th Lunar Planetary Science Conference, 2043–2058, 1980.

Burns J. A., Original evolution. *Planetary Satellites*, Eds.: J. A. Burns, 113–156, Univ. of Arizona, Tucson, 1977

Cameron A. G. W., The origin of the Moon and single impact hypothesis. *Icarus*, **126**: 126–137, 1997.

Cameron A. G. W., Higher-resolution simulations of the giant impact. Origin of the. Earth and Moon Eds.: K. Righter & R. Canup, 133–144, Lunar Planet. Inst., Houston, 2000.

Cameron A. G. W. and Ward W., The Origin of the Moon. In: Proceedings of 7th Lunar Science Conference, 120–122, 1976.

Campbell I. H. and O'Neil, H. C., Evidence against a chondritic Earth. *Nature* **483**: 553–558, 2012.

Canup R. M., Simulations of a Late Lunar Forming Impact. *Icarus* **168**: 433–456, 2004

Canup R. M., Lunar forming collisions with preimpact rotation. *Icarus* **196**: 518–538, 2008.

Canup R. M. and Asphaug E., Origin of the Moon in a Giant Impact near the End of the Earth's Formation, *Nature* **41**(6848): 708–712, 2001.

Canup R. M. and Esposito L. W., Formation of the Moon from an impact-generated disk. *Icarus*, **119**: 427–446, 1996.

Canup R. M., Implication of lunar origin via impact for the Moon's composition and thermal state of the protoearth. In: Proc. 39th Lunar Planet. Sci. Conf., abst. 2429, 2008.

Canup R. M. and Ward W. R., Formation of the Galilean satellites: conditions of accretion, *Astron. J* **124**: 3404–3423, 2002.

Capobianco C. J., Jones J. H. and Drake M. J., Metal-silicate thermochemistry at high temperature: magma ocean and the "excess siderofile element" problem of the Earth's upper mantle. *L. Geophys. Res.* **98**(E3): 5433–5443, 1993.

Capuzzo-Dolcetta R. and Miocchi P., A Comparison between Fast Multipole Algorithm and Tree–Code to Evaluate Gravitational Forces in 3-D, astro-ph/9703122, **1**, 1997.

Carlson R. W. and Lugmair G. W., Time and duration of lunar highlands crust formation. *Earth Planet. Sci. Lett.* **52**: 227–238, 1981.

Carlson R. W. and Lugmair G. W., The age of ferroan anorthosite 60025: Oldest crust on a young Moon? *Earth Planet. Sci. Lett.* **90**: 119–130, 1988.

Carlson R. W. and Lugmair G. W., Timescales of planetesimal formation and differentiation based on extinct and extant radioisotopes. In: *Origins of the Earth and Moon* Eds.: R. M. Canup, K. Righter, 25–44, Univ. Arizona Press, 2000.

Cassen P., Guillot T. and Quirrenbach A., Extasolar Planets. Saas-Fee Advanced Course **31**: 450, Springer-Verlag, 2006.

Chabot N. L. and Agee C. B., Core formation in the Earth and Moon: new experimental constraints from V, Cr, and Mn. *Geochim. Cosmochim. Acta* **67**: 2077–2091, 2003.

Chabot N. L., Draper D. S. and Agee C. B., Conditions of core formation in the earth: Constraints from nickel and cobalt partitioning. *Geochim. Cosmochim. Acta* **69**: 2141–2151, 2005.

Chaktabarti R. and Jacobsen S. B., The isotopic composition of magnesium in bulk lunar soils. In: 42nd Lunar and Planetary Science Conference, Woodlands, USA, March 7-11, 2006, 2011.

Chambers J. E., Planetary accretion in the inner solar system. *Earth Planet. Sci. Lett.* **223**: 241–252, 2004.

Chen Y. and Zhang, Initial water concentration and degassing of lunar basalts inferred from melt inclusions in olivine. In: Proc 43rd Lunar and Planet Sci. Conf. Abst. 1361, 2012.

Clark R. N., Detection of absorbed water and hydroxyl on the Moon. *Science* **326**(5952): 562–564, 2009.

Clayton R. N., Oxygen isotopes in meteorites. In: Annu. Rev. *Earth Planet. Sci. Lett.* **21**: 115–149, 1993.

Clayton R. N., Solar System: self-shielding in the solar nebula. *Nature* **415**: 860–861, 2002.

Cisowski S. M. and Fuller M., Lunar paleointensities via the IRMs normalization method and the Early magnetic history of the Moon. In: *Origin of the Moon*, Eds.: W. K. Hartmann et al., 411–424, Lunar and Planet. Institute, Houston, 1986.

Colaprete A., et al., Detection of water in the LCROSS ejection plume, *Science* **330**: 463, 2010.

Cottrell E., Walter M. J. and Walker D., Metal-silicate partitioning of tungsten at high pressure and temperature: implications for equilibrium core formation in Earth. *Earth Planet Sci. Lett.* **281**: 275–287, 2009.

Cumming A., Marcy G. W., Butler B. P. and Vogt S. S., The statistics of extrasolar planets: results from the Keck survey. In: *Scientific frontiers in Research on Extrasolar Planets*, Eds.: D. Deming and S. Seager, ASP Conf. Ser. **294**: 27–30, 2002.

D'Angelo G., Henning T. and Kley W., Nested-grid calculations of disk-planet interaction. *A&A* **385**: 647, 2002.

Darwin G. H., On the precession of a viscous spheroid and on the remote history of the earth. *Phil. Trans. Roy. Soc. London*, **170**: 447–530, 1879.

Darwin G. H., On the presentation of a viscous spheroid. *Nature* **18**: 580–582, 1978.

Davis A. M., Hashimoto R. N., Clayton R. N. and Mayeda T. K., Correlated isotopic mass fractionation of oxygen, magnesium and silicon in forsterite evaporation residues. *Nature* **347**: 655–658, 1990.

Davies G. F. and Richards M. A., Mantle convection, *J. Geol.* **100**: 151–206, 1992.

Day J. M. D. and Walker R. J., The lunar mantle. In: 42nd Lunar Plan. Sci. Conference, abstr. 1288, 2011.

Delano J. W., Abundances of Cobalt, Nickel, and Volatiles in the Silicate Portion of the Moon.. In: *Origin of the Moon*, Eds. W. K. Hartmann, R. J. Phillips, and G. J. Taylor, 231–248, Lunar Planet. Inst., Houston, 1986.

De Maria G., Balducci G., Guido M. and Piacente V., Mass Spectrometric Investigation of the Vaporization Process of Apollo 12 Lunar Samples. In: Proc. 2nd Lunar. Conf. **2**: 1367–1380, 1971.

Descartes R., Le Monde, ou Traité de la Lumiére, 1664. Reprinted in Descartes, **4**: 215–332. 1824.

Dominguez G., Jackson T., Nunn M., Basov D. and Thiemens M., Low temperature mass-independent ozone formation on cold surfaces. In: Proc. 43rd Lunar and Planet. Scie. Conf., abst. 2403, 2012.

Dones L. and Tremaine S., Why does the Earth spin forward? *Science* **295**: 350–354, 1993.

Drake M. J., Is lunar bulk material similar to the Earth's mantle. In: *Origin of the Moon*, Eds. W. K. Hartmann, R. J. Phillips, G. J. Taylor., 105–124, Lunar Planet. Inst., Houston, 1986.

Edmunson J., Borg L. E., Nyquist L. E. and Asmerom Y., A combined Sm Nd, Rb Sr, and U Pb isotopic study of Mg suite norite 78238: Further evidence for early differentiation of the Moon. Lawrence Livermore National lab. JRNL-408877, Nov 19, Geochim. Cosmochim. Acta 2008.

Efroimsky M., Bodily tides near spin-orbit resonances. *Celest. Mech. Dyn. Astr.* **112**: 283–330, 2012.

Eke V. R., Teodoro L. F. A., Lawrence D. J., Elphic R. C. and Feldman W. C., What is the LEND collimated detector really measuring? In: Proc. 43rd Lunar and Planet Sci. Conf. Abst. 2211, 2012.

Elkins-Tanton L. T., Van Or man J. A., Hager B. H. and Grove T. L., Re-examination of the lunar magma ocean cumulate overturn hypothesis: melting or mixing is required. *Earth Planet. Sci. Lett.* **196**: 239–249, 2002.

Elkins-Tanton L. T., Burgess S. and Yin Q. Z., The Lunar Magma Ocean: Reconciling the Solidification Process with Lunar Petrology and Geochronology. In: Proc. 42nd Lunar and Planet. Scie. Confer., abst. 1505, 2011.

Eneev T. M., New Accumulation Model of Planet Formation and the Structure of the Outer Areas of the Solar System. Preprint No. 166, Inst. Prikl. Matem. Akad. Nauk SSSR, Moscow (in Russian), 1979.

Eneev T. M. and Kozlov N. N. Preprint No. 134, Inst. Prikl. Matem. Akad. Nauk SSSR, Moscow (in Russian), 1977.

Esat T. M. and Taylor S. R., Magnesium isotope fractionation in lunar soils. *Geochim. Cosmochim. Acta* **56**: 1025–1031, 1992.

Epstein S., and Taylor H. P., Stable isotopes, rare gases, solar wind, and spallation products. $^{18}O/^{16}O$, $^{30}Si/^{28}Si$, D/H, $^{13}C/^{12}C$ studies of lunar rocks and minerals. *Science* **167**: 533–535, 1970.

Fei L., Wenrui W., Wn C. and Weifeng H., Lunar global crustal thickness estimation using compensated terrain gravity effect (CTGE) data. In: Proc. 43rd Lunar and Planet. Sci. Confer., abst. 1432, 2012.

Feldman W. C., Maurice S., Binder A. B., Barraclough B. L., Elphic R. C. and Lawrence D. J., Fluxes of fast epithermal neutrons from Lunar Prospector: Evidence for water ice at the Lunar Poles. *Science* **281**: 1496–1500, 1998.

Fleet M. E., Stone W. E. and Crocket J. H., Partitioning of palladium, iridium and platinum between sulfide liquid and basalt melt: effect of the melt composition, concentration, and oxygen fugacity. *Geochim. Cosmoch. Acta* **55**: 2545–2554, 1991.

Friedman I. et al., Apollo 15 Lunar Samples, 302–306, 1972.

Fridman A. M. and Polyachenko V. L., *Physics of Gravitating Systems*. Springer Verlag, New York, Berlin, Heidelberg, Tokyo, **1**, 468 p., **2**, 358 p., 1984.

Frost D. J., Liebske Ch., Langenhorst F., McCammon C. A., Tronnes R. and Ruble D. C., Experimental evidence for the existence of iron-rich metal in the Earth's lower mantle. *Nature* **428**: 409–412, 2004.

Gaffney A. M., Borg L. E. and Asmeron Y., U-Pb systematics in mare basalt source regions: A combined U-Pb, Rb-Sr. and Sm-Nd study of mare basalts 10017. *Geochim.Cosmochim. Acta* **71**: 3656–3671, 2007.

Galimov E. M., Nuclear-spin-dependent isotope effect is a new type of isotope effects. *Geokhimiya*, **2**: 274–284, 1979.

Galimov E. M., Isotope fractionation related to kimberlite magmatism and diamond formation. *Geoch. Cosmoch. Acta* **55**(6): 1697–1708, 1991.

Galimov E. M., Problem of the Origin of the Moon. In: *Main Branches in Geochemistry*, Ed. E. M. Galimov, 8–45 (in Russian), Nauka, Moscow, 1995.

Galimov E. M., Growth of the Earth's core as a source of its internal energy and a factor of mantle redox evolution. *Geochem. Int.* **36**(8): 673–675, 1998.

Galimov E. M., On the origin of lunar material. *Geochem. Int.* **42**(7): 595–609, 2004.

Galimov E. M., Redox evolution of the Earth caused by a multi-stage formation of its core. *Earth Planet. Sci. Lett.* **233**: 263–276, 2005.

Galimov E. M., Formation of the Moon and the Earth from a Common Supraplanetary Gas Dust Cloud. Lecture Presented at the XIX All-Russia Symposium on Isotope Geochemistry on November 16, 2010. *Geochem. Int.* **49**(6): 537–554, 2011.

Galimov E. M. and Krivtsov A. M., Origin of the Earth-Moon system. *J. Earth Syst. Sci.* **114**(6): 593–600, 2005.

Galimov E. M., Krivtsov A. M., Zabrodin A. V., Legkostupov M. S., Eneev T. M. and Sidorov Yu. I., Dynamic Model for the Formation of the Earth-Moon System, *Geochem. Int.* **43**(11): 1045–1055, 2005.

Galimov E. M. and Ryzhenko B. N., Solution of the K/Na Biogeochemical Paradox. *Dokl. Earth Sciences* **421**(2): 911–913, 2008.

Galimov E. M., Ryzhenko B. N. and Cherkasova E. V., Reconstruction of chemical composition of water phase on the early Earth: 1. Formation from material of carbonaceous chondrites. *Geochem Int.* (6): 563–569, 2011.

Georg R. B., Halliday A. N., Schauble E. A. and Reynolds B. C., Silicon in the Earth's core. *Nature* **447**: 1102–1106, 2007.

Gessmann C. K. and Rubie D. C., The effect of temperature on the partitioning of nickel, cobalt, manganese, chromium, and vanadium at 9 GPa and constraints on formation of the Earth's core. *Geochim. Cosmochim. Acta* **62**: 867–878, 1998.

Gessmann C. K., Wood B. J., Rubie D. C. and Kilburn M. R., Solubility of silicon in liquid metal at high pressure: implications for the composition of the Earth's core. *Earth Planet. Sci. Lett.* **184**: 367–376, 2001.

Glushak B. L., Kuropatenko V. F. and Novikov S. A., *Investigation of material strength under dynamical loading*. Nauka, Novosibirsk, 295 pp. (in Russian), 1992.

Goins N. R., Dainty A. M. and Toksoz M. N., Lunar seismology: the internal structure of the Moon. *J. Geophys. Res.* **86**: 5061–74, 1981.

Goldreich P. and Ward W. R., The Formation of Planetesimals. *Astrophys. J.* **183**: 1057–1061, 1973.

Greenwood J. P. et al, Water in Apollo rock samples and the D/H of lunar Apatite. In: Proc. 41st Lunar and Planet Sci. Conf. Abst. 2439, 2010.

Greenwood J. P., Itoh S., Sakamoto N., Werren P. H., Dyar M. D. and Yurimoto H., Origin of Lunar water and evidence for a wet Moon from D/H and water in lunar Apatites. In: Proc 42nd Lunar and Planet Sci. Conf., Abst. 2753, 2011.

Grevese N. and Sauval A. J., Standard Solar Composition. In: Solar Composition and Evolution – from Core to Corona, Eds. C. Frohlich, M. C. E. Huber and S. K. Solanki, Space Sci. Rev. **85**: 161–174, 1998.

Grossman L. and Larimer J. W., Early Chemical History of the Solar System. *Rev. Geophys. Sp. Phys.* **12**: 71–101, 1974.

Gurevich L. E. and Lebedinskii A. I., Formation of Planets. *Izv. Akad. Nauk SSSR, Ser. Fiz.* **14**(6): 765–775, 1950.

Haggerty S. E., Boyd F. R., Bell P. M., Finger L. W. and Bryan W. B., Opaque minerals and olivine in lavas and breccias from Mare Tranquilitatis. In: Proc. Apollo 11 Lunar Sci. Conf., 513–538, 1970.

Haisch K. E., Loda E. A. and Loda C. L., Disk frequencies and life-times in young clusters. *Astrophys. Jour.* **553**: L.153–L.156, 2001.

Halliday A. N., A young Moon-forming giant impact at 70–110 million years accompanied by late-stage mixing, core formation and degassing of the Earth. *Phil. Trans. Roy. Soc. Lond. A* **366**: 4163–4181, 2008.

Halliday A. N. and Lee D. C., Tungsten Isotopes and the Early Development of the Earth and Moon. *Geochim. Cosmochim. Acta* **63**: 4157–4179, 1999.

Hanan B. B. and Tilton G. R., 60025: relict of primitive lunar crust? *Earth Planet. Sci. Lett.* **84**: 15–21, 1987.

Hamada T., Narumi T., Yokota R., Yasuoka K., Nitadori K. and Taiji M., 42 TFlops hierarchical N-body simulations on GPUs with applications in both astrophysics and turbulence. In: Proceedings of the Conference on High Performance Computing Networking, Storage and Analysis, article 62, 1–12, 2009.

Hartmann W. K. and Davis D. R., Satellite-Sized Planets and Lunar Origin, *Icarus* **24**: 504–515, 1975.

Harris A. W., Satellite formation, II. *Icarus*. **34**: 128–145, 1978.

Harris A. W. and Kaula W. M., Co-accretional model of satellite formation. *Icarus* **24**: 516–524, 1975

Hashimoto A., Evaporation Metamorphism in the Early Solar Nebula – Evaporation Experiments on the Melt FeO–MgO–SiO_2–CaO–Al_2O_3 and Chemical Fractionations of Primitive Materials. *Geochem. J.* **17**: 111–145, 1983.

Heiken G. H., Vaniman D. and French B. M., *Lunar Sourcebook: A USER's guide to the Moon*. Cambridge Univ. Press, 736 pp., 1991.

Head J. W. and Wilson L., Lunar mare volcanism: Stratigraphy, eruption conditions, and the evolution of secondary crust. *Geochim. Cosmochim. Acta* **56**: 2155–2175, 1992.

Heinsinger H. and Head J. W., Ages and stratigraphy of mare basalts in Oceanus Procellarum, Mare Nubium, Mare Cognitum, and Mare Insularium. *Jour. Geophys. Res.* **108**(E7) 5065, 2003.

Hernquist L., Performance Characteristics of Tree Codes. *Astrophys. J. Suppl. Ser.* **64**:715–734, 1987.

Hess P. C. and Parmentier E. M., A model for the thermal and chemical evolution of the Moon's interior: Implications from the onset of mare volcanism. *Earth Planet Sci. Lett.* **134**: 501–514, 1995.

Hockney R. W. and Eastwood J. W., *Computer Simulation using Particles*. Inst. Phys. Adam Hilger, Bristol. 561 p., 1988.

Holley E., Some account of the ancient state of the Cirty of Palmyra, with short remarks upon the inscriptions found there. *Phil. Trans. Roy. Soc. London* **19**: 160–175, 1695.

Holzheid A., Schmitz M. D. and Grove T. L., Textural equilibria of iron sulfide liquids in partly molten silicate aggregates and their relevance to core formation scenarios. *Jour. Geophys. Res.* **105**(B6): 13,555–13,567, 2000.

Hood L. L., Geophysical constraints on the lunar interior. In: *Origin of the Moon*, Eds: W. K. Hartman et al., 361–410, Lunar and Planet. Inst., Houston, 1986.

Hood L. L., Mitchell D. L., Lin R. P., Acuna M. N. and Binder A. B., Initial measurements of the lunar induced magnetic dipole moment using Lunar Prospector magnetometer data. *Geophys. Res. Lett.* **26**: 2327–2330, 1999.

Hood L. L. and Zuber M. T., Recent refinements in geophysical constrains on lunar origin and evolution. In: *Origin of the Earth and Moon*, Eds.: R. M. Canup and K. Righter, 397–409, Lunar and Planet. Inst., Houston, 2000.

Horan M. F., Smoltar M. L. and Walker R. J., W-182 and Re-187-Os-187 Systematics of Iron Meteorites: Chronology for Melting, Differentiation, and Crystallization in Asteroids. *Geochim. Cosmochim. Acta* **62**: 545–554, 1998.

Horedt G. P., *Polytropes: Applications in Astrophysics and Related Fields*. Kluwer Academic Publishers. 718 p., 2004.

Humayun M. and Cassen P., Processes Determining the Volatile Abundances of the Meteorites and Terrestrial Planets. In: *Origin of the Earth and Moon*, Eds.: R. M. Canup and K. Righter, 3–23, Univ. of Arizona Press, Tucson, 2000.

Humayun M. and Clayton R. N., Precise Determination of the Isotopic Composition of Potassium: Application to Terrestrial Rocks and Lunar Soils. *Geochim. Cosmochim. Acta* **59**: 2115–2130, 1995.

Hunten D. M., Pepin R. O. and Walker J. C. G., Mass Fractionation in Hydrodynamic Escape, *Icarus* **69**: 532–549, 1987

Hunter C., *Mon. Not. Roy. Astro. Soc.*, **126**(4): 299, 1963.

Ishibara Y. et al., *Geophys. Res. Lett.* **38**: LO3201, 2011.

Jana D. and Walker D., The influence of sulfur on partitioning of siderophile elements. *Geochim. Cosmochim. Acta* **61**(24): 5255–5277, 1996.

Jeffreys H. C., The resonance theory of the origin of the Moon (second paper). *Mon. Not. Roy. Astr. Soc.* **91**: 169–173, 1930.

Jones J. H. and Delano J. W., Three Component Model for the Bulk Composition of the Moon. *Geochim. Cosmochim. Acta* **53**: 513–527, 1984.

Jones J. H. and Palme H., Geochemical constraints on the origin of the Earth and Moon. In: *Origin of the Earth and Moon.* Eds.: R. Canup and K. Righter, Univ. Arizona Press, 197–216, 2000.

Jutzi M. and Asphaug E., The lunar far-side highlands as the late accretion of the Moon's companion. In: Proc. 42nd Lunar and Planrt. Sci. Confer., abst. 2126, 2011.

Kadik A. A., Lebedev E. B., Dorfman A. M. and Bagdasarov N. Sh., Separation of magmatic melt of crystals by use of high-temperature centrifugation. *Geokhimiya* **1**: 43–54, 1989.

Kant I., Die Frage: Ob die Erde veralte? Physikalisch erwogen, 1754. Reprinted in Kant, **1**: 193–213, 1910.

Karato S. I. and Spetzler H. A., Defect Microdynamics in Minerals and Solid-State Mechanisms of Seismic Wave Attenuation and Velocity Dispersion in the Mantle. *Rev. Geophys.* **28**: 399–23, 1990.

Kasting J. F., Eggler D. H. and Raeburn S., Mantle redox evolution and the oxidation state of the Archean atmosphere. *J. Geol.* **101**: 245– 257, 1993.

Kazenas E. K. and Tsvetkov Yu. V., *Thermodynamics of Oxide Evaporation*. LKI, Moscow. (in Russian), 2008.

Kegler P., Holzheid A., Frest D. J., Rubie D. C., Dohmen R. and Plame H., New Ni and Co metal-silicate partitioning data and their relevance for an early terrestrial magma ocean. *Earth Planet Sci. Lett.* **268**: 28–40, 2008.

Khan A., Mosegaard K. and Rasmussen K. L., A new seismic velocity model for the Moon from a Monte Carlo inversion of the Apollo lunar seismic data. *Geophys. Res. Lett.* **27**: 1591–1594, 2000.

Khan A. and Mosegaard K., New information on the deep lunar interior from an inversion of lunar free oscillation periods, *Geophys. Res. Lett.* **28**:1791–94, 2001.

Kleine T., Palme H., Mezger K. and Halliday A. N., Hf–W Chronology of Lunar Metal and the Age and Early Differentiation of the Moon. *Science* **310**: 1671–1674, 2005.

Kleine T., Touboul M., Van Orman J. A., Bourdon B., Maden C., Mezger K. and Halliday A. N., Hf-W thermochronometry: Closure temperature and constraints on the accretion and cooling history of H chondrites parent body. *Earth Planet. Sci. Letters* **270**: 106–118, 2008.

Kleine T., Touboul M., Bourdon B., Nimmo F., Mezger K., Palme N., Jacobsen S., Yin Q.-Z. and Halliday A. N., Hf-W chronology of the accretion and early evolution of asteroids and terrestrial planets. *Geochim. Cosmochim. Acta* **73**: 5150–5188, 2009.

Konopliv A. S. et al., Improved gravity field of the Moon from Lunar Prospector. *Science* **281**: 1476–1480, 1998.

Konopliv A. S. et al., Recent gravity models as a result of the Lunar Prospector mission. *Icarus* **150**: 1–18, 2001.

Kostitsyn Yu. A., Terrestrial and chondritic Sm-Nd and Lu-Hf isotopic systems: Are they identical? *Petrolog. Intern.* **12**(5): 451–466, 2004.

Kramers J. D., Reconciling siderophile element data in the Earth and Moon, W isotopes and the upper lunar age limit in a simple model of homogenous accretion. *Chem. Geol.* **145**: 461–478, 1998.

Krivtsov A. M., Deformation and Fracture of Solids with Microstructure. Moscow, Fismatlit, 304 pp. (in Russian), 2007.

Kurahashi E., Kita N. T., Nagahara H. and Morishita Y., ^{26}Al-^{26}Mg systematics and petrological study of chondrules in CR chondrites. In: Goldschmidt Conference Abstracts, July 13–18, Vancouver, Canada, A504, 2008.

Kuskov O. L. and Fabrichnaya O. B., Constitution of the Moon: 2. Composition and Seismic Properties of the Lower Mantle. *Phys. Earth Planet. Int.* **83**: 197–216, 1994.

Kuskov O. L. and Kronrod V. A., Constitution of the Moon 5: Constraints on composition, density, and radius of a core. *Phys. Earth Planet Intern.*, **70**: 285–306, 1998.

Kuskov O. L., Kronrod V. A. and Hood L. L., Geochemical constraints on the seismic properties of the lunar mantle. *Phys. Earth Planet Intern.* **134**: 175–189, 2002

Lambec K. and Pullan S., The lunar fossil bulge hypothesis revisited. *Phys. Earth. Planet. Inter.* **22**: 29–35, 1980.

Langseth M. G., Keihm S. J. and Peters K., Revised lunar heat-flow values. In: Proc. 7th Lunar Sci. Conf, 3143–3171, 1976.

Laplace P. S., *Exposition du système du monde*. Paris, 1796.

Larimer J. W., The Condensation and Fractionation of Refractory Lithophile Elements. *Icarus* **40**: 446–454, 1979.

Larimer J. W., Nebular Chemistry and Theories of Lunar Origin. In: *Origin of the Moon*, Eds.: W. K. Hartman, R. J. Phillips, and G. J. Taylor, 145–171, Lunar Planet. Inst., Houston, 1986.

Laskar J. and Robutel P., The chaoticobliquity of the planets. *Nature* **361**: 608–612, 1993.

Lebedev E. B. and Galimov E.M., Experimental modelling of the origin of the Moon's core under conditions of partial melting. *Geokhimiya*, **51**(8): 715–725, 2012.

Lebedev E. B., Kadik A. A., Kuskov O. L., Dorfman A. M. and Lukanin O. A., Sulphide phase moving in partially melted silicates. Implications to formation of planetary cores. *Astronomical News* **33**: 395–405, 1999.

Lecuyer Ch. and Ricard Ya., Long-term fluxes and budget of ferric iron: implication for the redox states of the Earth's mantle and atmosphere. *Earth Planet. Sci. Lett.* **165**: 197–211, 1999.

Lee D.-C. and Halliday A. N., Hafnium-Tungsten Chronometry and the Timing of Terrestrial Core Formation. *Nature* **378**: 771–774, 1995.

Lee D.-C., Halliday A. N., Snyder G. A. and Taylor L. A., Age and Origin of the Moon. *Science* **278**: 1098–1103, 1997.

Legostaev V. P. and Lopota V. A. Editors, and 60 contributors, *The Moon – step to exploration and utilization tehnologies of the solar system*, Energia, 584 p. (in Russian), 2011.

Leya I., Rainer W. and Halliday A. N., Cosmic-Ray Production of Tungsten Isotopes in Lunar Samples and Meteorites and Its Implications for Hf–W Cosmochemistry. *Earth Planet. Sci. Lett.* **175**: 1–12, 2000.

Le-Zaharov A. A., Volkovets I. B. and Krivtsov A. M., Parallel implementation of Barnes-Hut algorithm for simulation of planet system formation. In: Proc. of XXXIII Summer School Advanced Problems in Mechanics 2005, 237–242, St. Petersburg, Russia, 2006.

Le-Zakharov A. A. and Krivtsov A. M., Development of algorithms for the collisional dynamics computations of gravitating particles for simulation of the Earth-moon system formation as a result of the gravitational collapse of a dust cloud. In: Report for the project Problems of biosphere origin and evolution, Russian Academy of Sciences (in Russian). To appear in: *Problems of biosphere origin*, URSS 2012; 2009.

Le-Zakharov A. A. and Krivtsov A. M., Development of algorithms for computing the collisional dynamics of gravitating particles to simulate the formation of the earth-moon system through the gravitational collapse of a dust cloud. In: *Problems of biosphere origin and evolution*. Ed.: Eric M. Galimov, 317–329, Nova Science Publishers, NY., 2012.

Li J. and Agee C. B., Geochemistry of mantle-core differentiation at high pressures. *Nature* **381**: 686–689, 1996.

Lillis R. J., Frey H. V. and Manga M., Rapid decrease in Martian crustal magnetization in the Noachian era: Implications for the dynamo and climate of early Mars. *Geophys. Res. Lett.* **35**: L14203, doi:10.1029/2008GL034338, 2008.

Lin R. P., Mitchell D. L., Curtis D. W., Anderson K. A., Carlson C. W., McFadden J., Acuna M. H., Hood L. L. and Binder A., Lunar surface magnetic fields and their interaction with the Solar wind: Results from Lunar Prospector. *Science* **281**: 1480–1484, 1998.

Lissauer J. J., Dones L. and Ohtsuki K., Origin and evolution of terrestrial planet rotation. In: *Origin of Earth and Moon*, Eds.: R. M. Canup and R. Righter, 101–112, University of Arisona Press, Tucson, 2000.

Liu Y. Guan Y., Zhang Y., Rossman G. R., Eiler J. M. and Taylor L. A., Lunar surface water in agglutinates: origin and abundances. In: Proc. 43rd Lunar and Planet Sci. Conf. abst. 1864, 2012.

Lognonne Ph., Gagnepain-Beyneix J. and Chenet H., A new seismic model of the Moon: implication for structure, thermal evolution and formation of the Moon. *Earth Planet. Sci. Lett.* **211**: 27–44, 2003.

Lognonne Ph., Planetary seismology. *Ann. Rev. Earth Planet. Sci.* **33**: 19.1–19.34, 2005.

Longhi J., Petrogenesis of picritic mare magmas: Constraints on the extent of early lunar differentiation. *Geochi. Cosmochim. Acta* **70**: 5919–5934, 2006.

Lyons J. R. and Young E. D., CO self-shielding as the origin of oxygen isotope anomalies in the early solar nebula. *Nature* **435**: 317–320, 2005.

Magna T., Wiechert U. and Halliday A. N., New constraints on the lithium isotope compositions of the Moon and terrestrial planets, *Earth Planet Sci. Lett.* **243**: 336–353, 2006.

Makino J., A fast parallel treecode with GRAPE. In: Publications of the Astronomical Society of Japan, **56**: 521–531, 2004.

Markova O. M., Yakovlev O. I., Semenov G. L. and Belov A. N., Some General Experimental Results on Natural Melt Evaporation in the Knudsen Cell. *Geokhimiya* **11**: 1559–1569, 1986.

Marov M. J., Kolesnichenko A. V., *Turbulence and Self-Organization. Problems of Space and Environments Modeling. Springer.* In press. (Russian version published 2009, URSS, 632 p.), 2012.

Mayer J. R., Beiträge zur Dynamik des Himmelts in populärer Darstellung. Verlag Ulfich Landherr, Heilbronn, 1848. English translation in Lindsay, 148–196, 1973.

McCubbin F. M., Shearer C. K. and Sharp Z. D., Magmatic volatiles in lunar apatite: Approaching a single solution to many unique observations. In: Proc. 42nd Lunar and Planet. Sci. Conf., abst. 2341, 2011.

McCulloch M. T., Primitive Sr-87/Sr-86 from an Archean barite and conjecture on the Earth's age and origin. *Earth Planet. Sci. Lett.* **126**(1-3): 1–13, 1994.

McDonald G. J. F., Origin of the Moon: Dynamical considerations. In: *The Earth-Moon System* Eds.: B. G. Marsden and A. G. Cameron, 165–209, Plenum, NY, 1966.

McDonough W. F., Compositional model for the Earth's core. In: *Treatise on Geochemistry* Ed. R. W. Carlson, v. **2**, Pergamon Press, Oxford, pp. 547–568, 2003.

McGovern P. J., An intrusive origin for lunar mascones: Magma ascent theory, gravitational signature and tests for GRAIL. In: Proc. 43rd Lunar and Planet. Sci. Conf., abst.2937, 2012.

McKeegan K. D., Kallio A. P. A., Heber V. S., Jarzebinski G., Mao P. H., Coath C. D., Kunihiro T., Wiens R. C., Nordholt J. E., Moses Jr. R. W., Reisenfeld D. B., Jurewicz A. J. G. and Burnett D. S., The Oxygen Isotopic Composition of the Sun Inferred from Captured Solar Wind. *Science*, **332**: 1528–1532, 2011.

McKeegan K. D. et al., Isotopic compositions of cometary matter returned by Stardust *Science* **314**: 1724–1728, 2006.

Melosh H. J., A New and Improved Equation of State for Impact Computations. In: Proc. 31st Lunar Planet. Conf., abst. 1903, 2000.

Melosh H. J. and Sonett C. R., When Worlds Collide: Jetted Vapor Plumes and the Moon's Origin. In: *Origin of the Moon*, Eds.: W. K. Hartmann, R. J. Phillips, and G. J. Taylor, 621–642, Lunar Planet. Inst., Houston, 1986.

Meyer J. et al., Coupled thermal-orbital evolution of the early Moon. *Icarus* **208**: 1–10, 2010.

Miller R. S., Nerurkar G. and Lawrence D. J., New insights into hydrogen at the lunar poles from detection of Fast and epithermal neutron signatures. In: Proc. 43rd Lunar and Planet. Sci. Conf., abst. 1538, 2012.

Mitrofanov I. G. et al., Hydrogen Mapping of the Lunar South Pole Using the LRO Neutron Detector Experiment LEND. *Science*, **330**: 438–448, 2010.

Mizuno H. and Boss A. P., Tidal disruption of dissipative planatesimals. *Icarus* **63**: 109–133, 1985.

Mukhin M. A. and Volkovets I. B., Investigation of parallelization algorithms for molecular dynamics problems. XIV Sacharov readings (in Russian), 2004.

Muller P. M. and Sjogren W. L., Mascons: Lunar Mass Concentrations. *Science* **161**: 680–684, 1968.

Nakamura Y., Seismic velocity structure of the lunar mantle. *J. Geophys. Res.* **88**: 677–686, 1983.

Nakajima M. and Stevenson D. J., The initial state of the Moon forming Disk and the Earth's Mantle Based on SPH Simulations. In: Proc. 43rd Lunar Planet. Sci. Conf., abst. 2627, 2012.

Namiki N., Iwata T., Matsumoto K., Hanada H., Noda H., Goossens S., Ogawa M., Kawano N., Asari K., Tsurata S., Ishihara Y., Lin Q., Kikuchi F., Ishikawa T., Sasaki S., Aoshima C., Korosawa K., Sugita S. and Takano T., Far-side gravity field of the Moon from four-way Doppler measurements of Selene (Kaguya). *Science* **323**: 900–904, 2009.

Neumann G. A. et al., The lunar crust: Global structure and signature of major basins. *J. Geophys. Res.* **101**: 16.841–16.863, 1996.

Newsom H. E. and Runcorn S. K., New Constraints of the Size of the Lunar Core and the Origin of the Moon. In: Proceedings of 22nd Lunar Planetary Science Conference., 973–974, 1991.

Norman M. D., Borg L. E., Nyquist L. E. and Bogard D. D., Chronology, geochemistry and petrology of a ferroan noritic anorthosite clast from Descartes breccia 67215: clues to the age, origin, structure, and impact history of the lunar crust. Meteorit. Planet. Sci. 38: 645–661, 2003.

Nozette S., Lichtenberg C. L., Spudis P., Bonner R., Ort W., Malaret E., Robinson M. and Shoemaker E. M., Clementine bi-static radar experiment: Preliminary resuits. Science 275(5292): 1495–1498, 1996.

Nyquist L. E., Reinold W. U., Bogard D. D., Wooden J. L., Bansal B. M., Weismann H. and Shih C. Y., A comparative Rb-Sr, Sm-Nd, and K-Ar study of shocked norite 78236: Evidence of show cooling of the lunar crust? In: Proc. 12th Lunar Planet. Sci. Conf., Houston, 69–97, 1981.

Nyquist L. E. and Shih C. Y., On the chronology of and isotopic record of lunar basaltic volcanism. Geochim.Cosmochim. Acta 56: 2213–2234, 1992.

O'Neil H. S., The origin of the Moon and the early history of the Earth. A chemical model, part 2, Geochim. Cosmochim. Acta 55: 1159–1172, 1991.

Örik E. J., Comments on lunar origin. Irish. Astron. Jour. 10: 190–238, 1972.

Ozima M., Noble gas state in the mantle. Rev. Geophys. 32: 405–426, 1994.

Pahlevan K. and Stevenson D., Equilibration in the aftermath of the lunar-forming giant impact. Earth Planet. Sci. Lett. 262: 438–449, 2007.

Palme H., Chemical and isotopic heterogeneity in protosolar matter. *Phil. Trans. Roy. Soc. Lond.* A, **359**: 2061–2075, 2001.

Palme H., Larimer J. W. and Lipschutz M. E., Moderately Volatile Elements. In: *Meteorites and the Early Solar System*, Eds: J. F. Kerridge and M. S. Matthews, 436–461, Univ. of Arizona Press, Tucson, 1998.

Papanastassiou D. A. and Wasserburg J. G., Initial strontium isotopic abundances and the resolution of small time differences in the formation of planetary objects. Contrib. Mineral. Petrol. 5: 361–365, 1969.

Pepin R. O. and Porcelli D., Xenon isotope sustematics, giant impact and mantle degassing on the early Earth. *Earth Planet. Sci. Lett.* **250**: 470–485, 2006.

Perera V. and Garrick-Bethell, Lunar asymmetry: Coincidence of the degree-1 and degree-2 features due to a Rayleigh-Taylor instability and reorientation. In: Proc. 42nd Lunar and Planet. Sci. Conf., abst. 2750, 2011.

Parmentier E. M. et al., Gravitational differentiation due to initial chemical stratification: origin of lunar asymmetry by the creep of dense KREEP? *Earth Planet. Sci. Lett.* **201**: 473–480, 2002.

Phillips R. J., Mascons: Progress toward a unique solution for mass distribution. *J. Geophys. Res.* **77**: 7106–7114, 1972.

Pieters C. M. et al., Character and spatial distribution of OH/H_2O on the surface of the Moon seen by M^3 on Chandrayaan-1. Science 326: 568–572, 2009.

Pinti D. L., Matsuda J. and Maruyama S., Anomalous Xenon in Archean cherts from Pilbara Craton. *Chem. Geol.* **174**: 387–395, 2001.

Podosek F. A. and Cassen P., Theoretical, observational, and isotopic estimates of the lifetime of the solar nebula. *Meteoritics* 29: 6–25, 1994.

Podosek F. A. and Ozima M., The Xenon Age of the Earth. In: *Origin of the Earth and Moon*. Ed.: R. M. Canap and K. Righter, 63–74, Univ. Arizona, 2000.

Poitrasson F., Does planetary differentiation really fractionate iron isotopes? *Earth and Plan. Sci. Lett.*, **256**, 484–492, 2007.

Poitrasson F., Halliday A. N., Lee D. C., Levasseur S. and Teutsch N., Iron isotope differences between Earth, Moon, Mars and Vesta as possible records of contrasted accretion mechanisms. *Earth Planet. Sci. Lett.* **223**: 253–266, 2004.

Porcelli D., Woollum D. and Cassen P., Deep Earth Rare Gases: Initial Inventories, Capture from the Solar Nebula, and Losses during Moon Formation. *Earth Planet. Sci. Lett.* **193**: 237–251, 2001.

Premo W. R. and Tatsumoto M., U-Th-Pb isotopic systematics of lunar norite 78235. In: Proc. 21st Lunar Planet. Sci. Conf., Houston, 89–100, 1991.

Premo W. R. and Tatsumoto M., U-Th-Pb and Sm-Nd isotopic systematics of lunar troctolitic cumulate 76535: Implications on the age and origin of this early lunar, deep-seated cumulate. In: Proc. 22nd Lunar Planet. Sci. Conf. 381–397, Lunar Planet. Inst., Houston, 1992.

Rai N. and van Westernen W., Constraints on the formation of a lunar core from metal-silicate partitioning of siderophile elements. In: Proc. 43rd Lunar and Planet. Sci. Conf., abst. 1781, 2012.

Reynolds J. H., Xenology. J. Geophys. Res., 68: 2939–2956, 1963.

Reufer A., Meier M. M. M., Benz W. and Wieler R., Obtaining Higher Target Material Proportions in the Giant Impact by Changing Impact Parameters and Impactor Composition. 42nd Lunar and Planetary Science Conference, abst. 1136, 2011.

Righter K. and Drake M. J., Metal/silicate equilibrium in the early earth: new constrains from the volatile moderately siderophile elements Ga, Cu, P and Sn. *Geochim. Cosmochim. Acta* **64**: 3581–3597, 2000.

Righter K., Drake M J. and Yaxley G., Prediction of siderophile element metal – silicate partition coefficient to 20GPa and 2800 °C: the effects of pressure, temperature, oxygen fugacity, and silicate and metallic composition. *Phys. Earth planet. Int.* **100**: 115–137, 1997.

Righter K., Humayun M., Campbell A. J., Danielson L. R. D. and Drake M. J., Experimental studies for the terrestrial and luner mantles. Geochim. Cosmochim. Acta 73: 1487–1504, 2009.

Righter K., Pando K. M., Danielson L. and Cin-Ty Lee, Partitioning of Mo, P and other sideraphile elements (Cu, Ga, Sn, Ni, Co, Cr, Mn, V and W) between metal and silicate melt as a function of temperature and silicate melt composition. *Earth Planet. Sci. Lett.*, **291**: 1–9, 2000.

Ringwood A. E., *Origin of the Earth and the Moon*. Springer-Verlag, NY, Heidelberg, Berlin, 1979.

Ringwood A. E., Composition and Origin of the Moon. In: *Origin of the Moon*, Eds.: W. K. Hartmann et al., 673–698, Lunar Planet. Inst., Houston, 1986.

Ringwood A. E. and Kesson, S. E., Basaltic magmatism and the bulk composition of the moon. II - Siderophile and volatile elements in moon, earth and chondrites: Implications for lunar origin. *Research School of Earth Sciences*, A.N.U., no. 1221-1222, 1976.

Ringwood A. E. and Kesson, S. E. Composition and origin of the Moon. 8th Lunar Science Conference, Houston, March 14-18, V. **1**. New York, Pergamon Press, 371-398, 1977.

Roberts J. J., Kinney J. H., Siebert J. and Ryerson F. J., Fe-Ni-S melt permeability in olivine: implication for planetary core formation. *Geophys. Res. Lett.* **34**: L14306. doi:10.1029/2007GL030497, 2007.

Robert F. et al., The Solar system D/H ratio: observations and theories. *Space Sci. Rev.* **92**: 201–224, 2000.

Rubie D. C., Melosh, H. J., Reid, J. E., Liebske, C. and Righter, K., Mechanisms of metal-silicate equilibration in the terrestrial magma ocean. *Earth Planet. Sci. Lett.* **205**: 239–255, 2003.

Runcorn S. K., The formation of the lunar core. *Geochim. Cosmochim. Acta* **60**: 1205–1208, 1996.

Rushmer T., An Experimental Deformation Study of Partially Molten Amphibolite: Application to Low-Melt Fraction Segregation. *J. Geophys. Res.* **100**: 15681–15695, 1995.

Rushmer T., Minarik W. G. and Taylor G. J., Physical Process of Core Formation, in *Origin of the Earth and Moon*, Eds. R. M. Canup and K. Righter, 227–243, Univ. of Arizona, Tucson, 2000.

Ruskol E. L., The origin of the Moon I. Formation of a swarm of bodies around the Earth. *Soviet Astronomy* AJ, **4**: 657–668, 1960.

Ruskol E. L., The origin of the Moon III. Some aspects of the dynamic of the circumterrestrial swarm of satellites. *Soviet Astronomy* AJ, **15**: 646–654, 1972.

Rutherford M. J. and Papale P., Origin of basalt fire-fountain eruptions on Earth versus the Moon. *Geology* **37**: 219–222, 2009.

Saal A. E., Hauri E. H., Van Orman J. A. and Ruberford M. J., D/H ratios of the Lunar volcanic glasses. In: Proc. 42nd Lunar and Planet Sci. Conf., abst. 1327, 2012.

Saal A. E. et al., Volatile content of the lunar volcanic glasses and the presence of water in the Moon's interior. *Nature* **454**: 192–195, 2008.

Safronov V. S., *Evolution of the Protoplanetary Cloud and Formation of the Earth and Planets*. Moscow, Nauka, 244 p. (in Russian), 1969.

Safronov V. S., *Selected works*, vol. 1: Origin of the Earth and Planets, 163–168 (in Russian), 2002.

Salmon J. J. and Canup R. M., Three-stage lunar accretion: Slow growth of the Moon and implications for Earth-Moon isotopic similarities. In: Proc. 43rd Lunar Planet. Sci. Conf., abst. 2540, 2012.

Sanin A. B., Mitrofanov I. G., Litvak M. L., Boynton W. V., Chin G. Droege G., Evans L. G., Garvin J., Golovin D. V., Harshman K., McClanahan T. P., Malakhov A., Mokrousov M. I., Milikh G., Sagdeev R. Z. and Star R. D., Testing of lunar permanently shadowed regions for water ice. Proc. 43rd Lunar and Planet Sci. Conf. Abst. 2134, 2012.

Sato M., The driving mechanism of lunar pyroclastic eruptions inferred from the oxygen fugacity behavior of Apollo 17 Orange Glass. In: Proc. 10th Lunar and Planet. Sci. Conf., 311–325, 1979.

Sedaghatpour F. S., Teng F.-Z., Liu Y., Sears D. W. G. and Taylor L. A., Behavior of Magnesium Isotopes During Lunar Magmatic Differentiation. In: Proc. 43rd Lunar Planet. Sci. Conf., abst. 2884, 2012).

Shannon M. C. and Agee C. B., Percolation of Core Melts at Lower Mantle Conditions. *Science* 15 May. **280**, 1998.

Sharpton V. L. and Head J. W., Stratigraphy and structural evolution of southern Mare Serenitatis. A reinterpretation based on Apollo lunar sounder experimental data. *J. Geophys. Res.* **87**: 10983–10998, 1982.

Schoenberg R., Kamber B. S., Collerson K. D. and Eugster O., New W-Isotope Evidence for Rapid Terrestrial Accretion and Very Early Core Formation. *Geochim. Cosmochim. Acta* **66**: 3151–3160, 2002.

Shih C.-Y., Nyquist L. E., Dasch E. J., Bogard D. D., Bansal B. M. and Wiesmann H., Age of pristine noritic clasts from lunar breccias 15445 and 15455. *Geochim. Cosmochim. Acta* **57**: 915–931, 1993.

Shukolykov Yu. A., Meshik Y. P., Jessberger E., Dang Vu Minh and Jordan J., Chemically fractionated fission xenon in meteorites and on the Earth. *Geochim. Cosmoch. Acta* **58**: 3075–3092 1994.

Shukolyukov A. and Lugmair G. W., On the 53Mn heterogeneity in the early solar system. *Space Sci. Rev.* **92**: 225–236, 2000.

Singer S. F., Origin of the Moon by capture. In: *Origin of the Moon*, Eds.: W. Hartman et al., 471–486, Lunar and Planet. Inst., Houston, 1986.

Smith D. E. et al., Topography of the Moon from the Clementine lidar. *J. Geophys. Res.* **102**: 1591–1611, 1997.

Sneade C. J., McKeegan K. D., Burchell M. J. and Kearsley A. T., Oxygen isotope measurements of simulated Wild 2 impact crater residues. In: Proc. 43rd Lunar and Planet. Scie. Confer., abst. 2238, 2012.

Snyder G. A., Borg L. E., Nyquist L. E. and Taylor L. A., Chronology and isotopic constraints on lunar evolution. In: *Origin of the Earth and Moon*, Eds. R. M. Canup and K. Righter, 361–395, Univ. Arizon Press, 2000.

Stevenson D. J., The nature of the earth prior to the oldest known rock record: the hadean Earth. In: *Earth's Earliest Biosphere: Its Origin and Evolution,* Ed.: J. W. Stopf, 32–40, Princeton Univ. Press, 1983.

Stevenson D. J., Origin of the Moon: the collision hypothesis. *Ann. Rev. Earth Planet. Sci.* **15**: 271–315, 1987.

Stevenson D. J., Fluid Dynamics of Core Formation, in *Origin of the Earth*, Eds. H. E. Newsom and J. H. Jones, 231–249, Oxford Univ. Press, 1990.

Stevenson D. J., Earth Formation: Combining Physical Models with Isotopic and Elemental Constraints. *Geochim. Cosmochim. Acta* 15th Goldshmidt Conference, abst A382, 2005.

Stewart G. R. and Kaula W. M., A Gravitational Kinetic Theory for Planetesimales. *Icarus* **24**: 516–524, 1980.

Suavet C., Weiss B. P., Fuller M. D., Gatteccecea J., Grove T. L. and Shuster D. L., Persistence of the linar dynamo units 3.6 billion years ago. In: Proc. 43rd Lunar Planet. Sci. Conf., abst. 1925, 2012.

Swindle T. D. and Podosek F. A., Iodine–Xenon Dating. In: *Meteorites and the Early Solar System*, Eds.: J. F. Kerridge and M. S. Matthews, 1127–1146, Univ. Arizona, Tucson, 1988.

Taylor L. A. et al., Lunar Mare Soils: Space weathering and the major effects of surface-correlated nanophase Fe. *J. Geophys. Res.* **106**: 27985–27999, 2001.

Taylor S. R., The Origin of the Moon: Geochemical Consideration. In: *Origin of the Moon"*, Eds. W. K. Hartmann, R. J. Phillips, and G. J. Taylor, 25–144, Lunar Planet. Inst., Houston, 1986.

Tera F., Papanastassiou D. A. and Wasserburg G. J., A lunar cataclysm at ∼ 3.95 AE and the structure of the lunar crust. In: Proc. 4th Lunar. Conf., Houston, 723–725, 1973.

Thiemens J. E. and Heidenreich J. E. III., The mass-independent fractionation of oxygen: A novel isotope effect and its possible cosmochemical implications. *Science* **219**: 1073–1075, 1983.

Tikoo S. M., Weiss B. P., Grove T. L., and Fuller M. D., Decline of the ancient lunar core dynamo. In: Proc. 43rd Lunar and Planet Sci. Confer., abst. 2691, 2012.

Toksöz M. N., Dainty A., Solomon S. C. and Anderson K., Structure of the Moon. *Reviews of Geophysics and Space Physics*, **12**: 539, 1974.

Tolstikhin I. N. and Kramers J. D., *The evolution of matter: from the Big Bang to the present day Earth*. Cambridge University Press. Cambridge UK, 532 pp., 2008.

Tompkins S. and Pieters C. M., Mineralogy of the lunar crust: results from Clementine. *Meteorit. Planet. Sci.* **34**: 25–41, 1999.

Touboul et al., Late formation and prolonged dierentiation of the Moon inferred from W isotopes in lunar metals. *Nature* 450, pp. 1206–1209, 2007.

Trinquier A., Elliot T., Ulfbeck D., Coath Ch., Krot F. H. and Bizzarro M., Origin of Nucleosynthetic Isotope Heterogeneity in the Solar Protoplanetary Disk. *Science*, **324**: 374–376, 2009.

Urey H. C., *The planets*. Yale Univ., New Haven, 245 pp., 1952.

Vasiliev S. V., Krivtsov A. M. and Galimov E. M., Study of the Planet-Satellite System Growth Process as a Result of the Accumulation of Dust Cloud Material. *Solar System Research* **45**(5): 410–419, 2011.

Verhoogen J., *Energetic of the Earth*, Nat. Acad. Sci, Washington, DC, 139 pp., 1980.

Vityazev A. V., Pechernikova G. V. and Safronov V. S., Terrestrial Planets: Origin and Early Evolution. *Nauka*, Moscow, (in Russian), 1990.

Wade J. and Wood B. J., Core formation and the oxidation state of the Earth. *Earth Planet. Sci. Lett.* **236**: 78–95, 2005.

Walter M. J., Newsome H. E., Ertel W. and Holzheid A., Siderophile elements in the Earth and Moon: Metal/Silicate partitioning and implication for core formation. In: *Origin of the Earth and Moon*, Eds.: R. Canup and K. Righter, Univ. Arizona Press, 265–290, 2000.

Walter M. J. and Thibault Y., Partitioning of tungsten and molybdenum between metallic liquid and silicate melt. *Science* **270**: 1186–1189, 1995.

Wang J., Davis A. M., Clayton R. N. and Mayeda T. K., Chemical and Isotopic Fractionation during the Evaporation of the FeO-MgO-SiO_2-CaO-Al_2O_3-TiO_2-REE Melt System. In: Proceedings of 30th Lunar Planetary Science Conference, 1459, 1994.

Wang J., Davis A. M, Clayton R. N. and Hashimoto A., Evaporation of Single Crystal Forsterite: Evaporation Kinetics, Magnesium Isotope Fractionation and Implication of Mass-Dependent Isotopic Fractionation of Mass-Controlled Reservoir. *Geochim. Cosmochim. Acta* **63**: 953–966, 1999.

Wang L., Davis A. M., Clayton R. N., Mayeda T. K. and Hashimoto A., Chemical and isotopic fractionation during the evaporation of the FeO-MgO-SiO_3-$CaOAl_2O_3$-TiO_2 rare earth element melt system. *Geochim. Cosmochim. Acta* **65**: 479–494, 2001.

Wänke H., Constitution of terrestrial planets. *Phil. Trans. Roy. Soc. Land.* **A393**: 287–302, 1981.

Wänke H. and Dreibus G., Geochemical Evidence for the Formation of the Moon by Impact-Induced Fission of the Proto-Earth. *Origin of the Moon*, Eds.: W. K. Hartman et al., 649–672, Lunar Planet. Inst., Houston, 1986.

Wänke H., Dreibus G. and Jagoutz E., Mantle chemistry and accretion history of the Earth. In: Kroner, A. (Ed.), *Archean Geochemistry*. Springer, Berlin, 1–24, 1984.

Warren P. H., The magma ocean concept and lunar evolution. Ann. Rev. Earth Planet. Sci. **13**: 201–240, 1985.

Warren P. H., Anorthosite Assimilation and the Origin of the Mg/Fe-Related Bimodality of Pristine Moon Rocks: Support form the Magmasphere Hypothesis. In: Proceedings of 16th Lunar Planetary Science Conference, pp. 331–343, 1986.

Warren P. H., Constraints on the impact-accreted capapace hypothesis for the lunar far side highlands. In: Proc. 43rd Lunar and Planetary Science Conference, abst. 2941, 2012.

Warren P. H., Tonui E. and Young E. D., Magnesium isotopes in lunar rocks and glasses and implications for the origin of the Moon. Lunar Planet. Sci. Conf. 36, abst. 2143, 2005.

Warren U., Inheritance of Magma Ocean Differentiation through Lunar Origin by Giant Impact. In: Proceedings of 22nd Lunar Planetary Science Conference, 1495–1496, 1992.

Watson K. et al., The behavior of volatiles on the lunar surface. *J. Geophys. Res.* **66**: 3033–3045, 1961.

Weber R. C. et al., Seismic Detection of the Lunar Core. *Science* **331**: 309–312, 2011.

Weber A., Saal A. E., Hauri E. H., Rutherford M. J. and Van Orman J., The volatile content and D/H ratios of the lunar picritic glass. In: Proc. 42nd Lunar and Planet Sci. Conf. Houston, Abst. 2571, 2011.

Weiczorek M. A., Jolliff B. L., Khan A., Pritchard M. E., Weiss B. P., Williams J. G., Hood L. L., Righter K., Neal C. R., Shearer C. K., McCallum I. S., Tompkins S., Hawke B. R., Peterson Ch., Gillis J. J. and Bussey B., The constitution and structure of the lunar interior. *Rev. Min. Geochem.* **60**: 221–364, 2006.

Weidenschilling S. J., Greeberg R., Chapman C. R., Herbert F., Davis D. R., Drake M. J., Jones J. and Hartmann W. K., Origin of the Moon from a circumterrestrial disk. In: *Origin of the Moon* Eds.: W. Hartmann et al., 731–762, Lunar and Planet. Inst., Houston, 1986.

Weitz C. M., Rutherford M. J. and Head J. W., Oxidation states and ascent history of the Apollo 17 volcanic beads as inferred from metal-glass equilibria, *Geochim. Cosmochim. Acta* **61**: 2765–2775, 1997.

Westrenen W., de Meijer R. J., Anisichkin V. F. and Voronin D. V., Forming the Moon from terrestrial silicate-rich material. In: 2012 edition Proc. 43rd Lunar and Planet. Sci. Conf., abst. 1738, 2012.

Wetherill G. W., Formation of the Terrestrial Planets, *Ann. Rev. Astron. Astrophys.* **18**: 77–113, 1980.

Wetherill G. W., Occurrence of Giant Impacts during the Growth of the Terrestrial Planets. *Science* **228**: 877–879, 1985.

Wetherill G. W. and Cox L. P., The Range of Validity of the Two-Body Approximation in Models of Terrestrial Planet Accumulation. *Icarus* **63**: 290–303, 1985.

Weyer S., Anbar A. D., Brey G. P., Münker C., Mezger K. and Woodland A. B., Iron isotope fractionation during planetary differentiation. *Earth Planet. Sci. Lett.*, **240**: 251–264, 2005.

White R. J. and Hillenbrand L. A., A Long-lived Accretion Disk around a Lithium-depleted Binary T Tauri Star. *Astrophys. J.* **621**(1): L65–L68, 2005.

Wiechert U. and Halliday A. N., Non-chondritic magnesium and the origins of the inner terrestrial planets. Earth Planet. Sci. Lett. 256(3-4): 360–371, 2007.

Wiechert U., Halliday A. N., Lee1 D. C., Snyder G. A., Taylor L. A. and Rumble D., Oxygen isotopes and the Moon-forming giant impact. *Science* **294**: 345–348, 2001.

Williams H. M., Marcowski A., Quitte G., Halliday A. N., Teutsch N. and Levasseur S., Fe isotope fractionation in iron meteorites: New insighte into metal-sulphide segregation and planetary accretion. *Earth Planet. Sci. Lett.* **250**: 486–500, 2006.

Williams J. G., Boggs D. H., Yoder C. F., Ratcliff J. T. and Dickey J. O., Lunar rotational dissipation in solid body and molten core. *J. Geophys. Res. Planets* **106**(E7): 27933–27968, 2001.

Williams J. G. and Dickey J. O., Lunar Geophysics, Geodesy, and Dynamics. In: Proceedings of 13th International Workshop on Laser Ranging, Eds. R. Noomen, S. Klosko, C. Noll, and M. Pearlman, 75–86, NASA/CP-2003-212248, 2003.

Williams J. G. and Boggs D. H., Lunar Core and Mantle. What Does LLR See? In: Proceedings of the 16th International Workshop on Laser Ranging, 12-17 October 2008, Poznan, Poland. Ed. S. Schilliak, 101–120, 2009.

Williams J. G., Boggs D. H. and Ratcliff J. T., Lunar moment of inertia. Love number and core. Proc. 43rd Lunar and Planet. Conf., abst. 2230, 2012.

Wilhelms D. E., The geologic history of the Moon. Washington. USGS Professional Paper **1342**: 205 p., 1987.

Wood J., Moon over Mauna Loa: A review of hypotheses of formation of Earth's Moon. In: *Origin of the Moon* Eds.: W. K. Hartmann, R. J. Phillips and G. J. Taylor, 17–56, Lunar and Planetary Institute, Houston, 1986.

Wood B. J., Wade J. and Kilburn M., Core formation and the oxidation state of the Earth: additional from Nb, V and Cr partitioning. *Geochim. Cosmochim. Acta* **72**: 1415–1426, 2008.

Yakovlev O. I., Kosolapov A. I., Kuznetsov A. V. and Nusinov M. D., The results of study of evaporation of basaltic melt in vacuum. *Dokl. Acad. Sci. USSR* **206**(4): (in Russian), (1972).

Yi W., Halliday A. N., Alt J. C., Lee D.-C., Rehkamper M., Garcia M. O. and Su Y. (2000) Cadmium, indium, tin, tellurium, and sulfur in oceanic basalts: implications for chalcophile element fractionation in the Earth. *J. Geophys. Res.* **105**(B8), 18927–18948.

Yin Q., Jacobsen S. B., Yamashita K., Blichert-Toff J., Telouk P. and Albarede F., A Short Timescale for Terrestrial Planet Formation from Hf–W Chronometry of Meteorites. *Nature* **418**: 949–952, 2002.

Youdin A. N. and Shu F. H., Planetesimal formation by gravitational instability. *Astrophys. J.* **580**: 494–505, 2002.

Young E. D., Assessing the implication or K isotope cosmochemistry for evaporation in the planetary solar nebula. *Earth Planet. Sci. Lett.*, **183**: 321–333, 2000.

Yurimoto H. and Kuramoto K., Molecular cloud origin for the oxygen isotope heterogeneity in the solar system. *Science* **305**: 1763–1766, 2004.

Zhang Y., Water in lunar basalts: the role of molecular hydrogen (H_2), especially in the diffusion of the H-component. In: Proc. 42nd Lunar and Planet. Scie. Conf. abst. 1957, 2011.

Zhang J., Danphas N. and Davis A. M., Titanium isotope homogeneity in the Earth-Moon system: evidence for complete isotope mixing between the impactor and the proto-Earth. 42nd Lunar and Planet. Scie. Conf. Woodlands, USA, March 7–11, abst. 1515, 2011.

Index

D/H 67, 68
δD 66–68
$\delta^{26}Mg$ 41
$\delta^{30}Si$ 39
$\delta^{7}Li$ 41
$\varepsilon^{53}Cr$ 40
$\varepsilon^{182}W$ 41
$\varepsilon^{46}Ti$ 39
$\varepsilon^{50}Ti$ 39
ε_w shift 80
^{129}I 81
$^{129}I-^{129}Xe$ 81
^{129}Xe 81
^{136}Xe 81
^{13}C isotope 63
$^{16}O/^{17}O/^{18}O$ 61
^{182}Hf 79
^{182}W 79
$^{182}W/^{184}W$ 41, 79
^{204}Pb 83
$^{206}Pb/^{204}Pb$ 32
$^{235}U-^{207}Pb$ 83
$^{235}U/^{204}Pb$ 84–86
$^{238}U-^{206}Pb$ 82, 83
$^{238}U/^{204}Pb$ 82, 84–86
$^{238}\mu$ 82, 83
^{244}Pu 81
$^{244}Pu-^{136}Xe$ 81
$^{46}Ti/^{47}Ti/^{50}Ti$ 61
$^{50}Ti/^{47}Ti$ 39
$^{53}Cr/^{52}Cr$ 40, 61
^{53}Mn 40
^{57}Fe 65
$^{87}Rb/^{86}Sr$ 87
$^{87}Sr/^{86}Sr$ 32

A
absolute dating 23
accomplishment of accretion 82

accretion 59
accretion of chondrites 47
accumulation 89
achondrites 31, 39, 40, 86
agglutinates 19, 68
alkali anorthosite 67
alkaline suite 21
aluminum contents 22
anarthositic crust 24
angular and random velocities 116
angular momentum 6, 43, 44, 49, 50
angular velocity 95, 96, 115
anorthite 21
anorthosite norites 21
anorthositic 27
apatite 21
Apollo missions 10
asteroids 7, 48
astronomic observations 42
asymmetry 14
augite 21
automatic sampling 12
average density 28
average lunar crustal 27

B
balancing constant 104
Barnes-Hut algorithm 94, 98, 107
bombardment 14

C
capturing 43
carbonaceous chondrite 31, 47, 53, 57, 80
carrier gas 57, 63
chondrites 32, 51, 80
chondrules 47
chromium 45
chromspinel 21
circular structures 17

circumterrestrial orbit 44
clinopyroxene 21
closed system 69
cloud 51, 57
co-accretion hypothesis 43
collapse 51
collapse dynamics 56
comet impacts 15
common source 49
compatible 73, 74
complete melting 76
computer simulation 56
condensation 33, 62, 63
condensation sequence 52
condensed bodies 51
condensed substances 52
conglomerations 51
continents 15
convection 36
core 29, 59, 65, 77, 80
core segregation 70, 82
cosmogenic neutrons 40
Cr-isotope composition 40
crater ejecta 17
Crisium 23
cristobalite 30
crust 14
crustal thickness 26
crystallization 20

D
decay 79
deficit in iron 44
degassing 59
degree of depletion 76
density 3, 10
depletion factor 74
deuterium 67, 69
diffusion 37, 67, 69
dimensionless parameter 50
diopside 21
dispersed 59
dissimilarity 14
dissipation coefficient 95, 115, 116
dissipative component 139
double Earth-Moon system 49
dunites 21
dust component 47

dust materials 55
dynamic equation 51
Dynamics 91
dynamics equations 49
dynamo 15

E
early atmosphere 30
Early hypotheses 42
Earth's core 44
Earth's mantle 44
Earth's rotation 42
elastic properties 27, 34
elevation dichotomy 14
embryo 57
embryo growth 125
embryo masses 135
embryos of Earth and Moon 55
energy dissipation 43, 50
enstatite 21
enstatite meteorite 40
equilibrium 69
equilibrium condensation 62
equilibrium partition 76
evaporation 32, 33, 37, 47, 49, 52, 53, 55, 57,
    62, 63, 69, 111
evaporative accretion 59–61, 63, 88
evolution 48
exposition 67

F
far side 10, 14
fast multipole method 99, 104
feeding cloud 137
feldspar 20
ferroan anorthosite 15, 20, 21, 25
ferrous silicates 25
fission 43
force calculation algorithm 94, 105
forsterite and enstatite 53
fractionation of isotopes 37
fragmentation 50, 55, 80, 88, 89, 93, 113, 125

G
gabbronorite 21
garnet 34
gas repulsion 138
gas-dust accumulation 89

Index

gas-dust cloud 56
gas-dust state 48
giant impact 48
giant impact concept 44
giant impact hypothesis 45, 80
gravitation 12
gravitation traps. 9
gravitational attraction 49
gravitational constant 94
gravitational field 13
gravitational instability 25
gravity anomalies 13
gravity attraction 8
gravity field 14, 25
gravity map 13
great bombardment 24

H
H_2O 68
Hafnium 79
halcophile elements 73
heating of particles 51
heating sources 36
Hf–W 79, 80
Hf/W 47, 79
high titanium basalts 24
highest elevation 15
highlands 10, 14
highly siderophile elements 71
Hill sphere 126, 139
homogenization 60
human landing 10
hydrated phase 66
hydrodynamic 32
hydrodynamic escape 59, 63, 82, 86
hydrodynamic flux 81, 82
hydrodynamic removal 57
hydrogen 57, 69
hydrogen abundance 65
hydrothermal alteration 47
hypothesis 43

I
ilmenite 21, 25
Imbrium 23
impact 44
impact basins 14
impact conditions 45

impact craters 48
impactor 45, 60, 61
inclination 5, 44
incompatible 72–74
indigenous lunar water 68
indigenous water 66
initial cloud density 115
initial Sr isotope ratio 85
initial Sr-isotope composition 86
interacting particles 49
interaction potential 116
internal lunar structure 25
interstellar 47
intrusions 21
inverse dissipation 27
iron 29, 44
iron content 29, 55
iron deficit 31
iron loss 51
iron oxide 53
isotope composition 60
isotope effects 64
isotope exchange 64
isotope fractionation 33, 41, 49, 61–64
isotope fractionation lines 68
isotope homogenization 60
isotopic composition 41, 63, 68
isotopic composition of hydrogen 66
isotopic composition of iron 64
isotopic composition of Pb 83
isotopic compositions of silicon 39
isotopic fractionation 45
isotopic ratios 60
isotopic shifts 33

J
Jupiter 10

K
Kepler's laws 8
kinetic isotope effect 62, 67
KREEP 24, 77, 82

L
Lagrange points 61
Lane-Emden equation 139
late veneer 70
lead 32

libration (Lagrangian) points. 8
lithophile 33
low titanium basalts 22
lower crust 21
lowlands 10
lunar basalts 67
lunar continents 19
lunar core 23
lunar core dynamo 15
lunar craters 10
lunar crust 15, 26, 34
lunar lithosphere 26
lunar magma ocean 25
lunar mountain 17
lunar orbit 4
lunar substance 69
lunar surface 66, 67
lunar topography 12
lunar water 61, 69

M
magma ocean 24, 82
magmatic cumulates 24
magmosphere 25
magnetic anomalies 15
magnetic field 15
mantle 29
mare basalts 22, 27, 30, 34
Mare Crisium 19
Mare Fecunditatis 19
Mare Imbrium 19
Mare Serenitatis 19
Mare Tranquillitatis 19
maria 15, 16, 19
maria basins 13
maria rocks 22
Mars 46
mascons 13, 14
mass ratio of embryos 127, 137
mass-independent isotope effects 39
Maturity of the regolith 19
maximum rotation rate 7
mega-impact hypothesis 45
melt inclusions 67
Mercury 46
merrillite 21
metal-melt equilibria 30
metal-silicate equilibrium 71

metallic core 30, 59, 69
metallic iron 19, 59, 64
metamorphosed matrix 47
Mg isotope composition 41
micro-meteorites 19
middle and lower mantle 34
minimum thickness 26
moderately volatile elements 52
moment of inertia 3
Multipole Acceptance Criterion 100, 103

N
Nectaris 23
nickel 29
norite 21
normalizing element 52
nucleosynthetic mechanism 37
number of clusters 116

O
Obliquity 5
Oceanus Procellarium 19
oldest lunar rocks 80
olivine 21, 30, 67
orange glass 30
orange glass balls 22
orbital evolution 7
ordinary chondrites 31, 39, 40, 59
orthopyroxene 21
overturn 25
oxidated state 30
oxygen 45, 68
oxygen isotope heterogeneity 38
oxygen isotopic 60
ozone 37

P
P-wave velocity 27
parallel computing 105
Parallelization scheme 104
partial melting 34, 74, 75, 78
particle collision 50
particle dynamics method 49, 93, 94
particle interaction 50
particle interaction force 94
partition 73
partition coefficient 70, 71, 73
Pb isotope ratios 82

percolation 77
percolation algorithm 141
percolation of metal 77
permeability 78
pigeonite 21
plagioclase 20, 21, 27
planet accumulation 44
planetesimals 43, 48
point of dynamic equilibrium 128
polar craters 65
polytropic 139
post-impact homogenization 45
potassium 33
potassium depletion 33
potassium feldspar 21
precision mapping 12
presence of water 12
primary matter 59
primary water 68
primordial 33
primordial atmosphere 81
primordial lunar material 77
prograde rotation 4
protoplanetary material 48
pyroclastic glasses 61
pyroclastic material 22
pyrolitic composition 27

Q

Q-factor 27
QFM buffer 30
quantity of heat 139
quartz 21
quartz monzodiorites 22

R

radioactive decay 35
radioactive heat 36
radioactive isotopes 33
Raleigh distillation 33
random velocity 115, 116
Rb-Sr 88
redox characteristics 30
redox conditions 30
redox evolution 36
reduced mantle 30
reducer 57
reduction of FeO 69

refractory elements 33, 34, 44, 49, 51, 52, 60, 73
regolith 18, 19, 66
repulsion 50
repulsive impulse 50
residual magnetization 15
rotation 4, 6
rotation axis 65
rotational collapse 50
rotational instability 112
rotational stability 7
Rubidium 32

S

sample return 12
scattered particles 55
screening 138
segregation 44, 77
seismic 34
seismic data 26
self-shielding 37
Serenitatis 23
shadowed craters 66
shergottites 20
siderophile distribution pattern 71
siderophile element 70–74, 76, 77
siderophile property 33, 73, 82
silica 44, 45
silicone isotope signature 39
similarity principle 96
soft landing 10
solar nebula 48, 57
solar system 60, 89
solar system formation 79, 80
solar wind 68
solid-body rotation 95, 96
solid-state-evolution 48
soviet spacecraft 10
spacecraft 66
spallation 67
Sr-isotope ratio 87, 88
stability 114
statistics of impact craters 23
stratification 20, 25
strong impact 27
successive condensation 63
superadiobatic temperature 36
surrounding material 57
synchronously 4

T

thermal energy 139
thermal state 35, 36
thermodynamic isotope effect 62–64
thickness of the crust 25
Ti isotope composition 40
tidal dissipation 35, 42
tidal heating 25
titanium content 22
topographies 14
tree traversal 101, 103
troctolite 21
Trojans 9
tungsten 45, 79

U

U–Pb system 84, 85
ultramafic rocks 21

undersaturated vapor phase 63
upper mantle 34

V

vaporization 62
Venus 46
volatiles 31, 45, 51, 59–61, 65, 69, 81, 82
volatility 33, 53, 69, 73
volcanic glasses 27, 66
volcanism 14

W

W isotopic compositions 79
water in lunar rocks 61
wetting 77

Z

zero-valency 19